Praise for *Extracted*

"Here is the book many of us in the sustainability world have been looking forward to: a comprehensive, readable, historically informed inquiry into the depletion of Earth's mineral resources. *Extracted* should be on the reading list of every introductory class in economics—as well as environmental studies, geology, history, political science . . . heck, everybody should read it."

—RICHARD HEINBERG, senior fellow,
Post Carbon Institute; author, *The End of Growth*

"The world economy is now phenomenally large in comparison with the planetary base that is the setting for all economic activity. Natural resources are becoming increasingly scarce, and the planet's sinks for absorbing waste products are already exhausted in many contexts. In *Extracted*, Ugo Bardi tells the story of our planetary plunder from its beginnings up through the present. He tells it with verve and insight, and he offers a powerful perspective on what the implications are for the future. This newest report from the Club of Rome demands our serious attention."

—JAMES GUSTAVE SPETH, author,
America the Possible: Manifesto for a New Economy;
former dean, Yale School of Forestry and Environmental Studies

"Most decision makers and citizens view money as the primary driver of our societies. Yet our civilization is first dependent on extraction of natural capital—minerals, ores, and particularly energy—that are the precursors for everything in our economies. Ugo Bardi and guest authors provide an excellent overview on the history, significance, and future of minerals and energy and how these relate to our human ecosystem. Wide-boundary thinking at its best."

—NATE HAGENS, *The Oil Drum*;
former vice-president, Salomon Brothers and Lehman Brothers

"Although Ugo Bardi's fine book focuses on extraction, it also discusses geological formation of minerals and ores, mining, metallurgy, coinage of precious metals, debt, waste, pollution, climate change, and the dark side of mining. Interspersed are short digressions written by other experts on related topics ranging from soil fertility and plants as miners, to peak oil and coal, and the Hubbert depletion curve. The book is clearly written and insightful. Highly recommended!"

—Herman Daly, author of *Ecological Economics*;
professor emeritus, School of Public Policy,
University of Maryland

"Ugo Bardi's book is an effective piece of work for stimulating thought and debate on this planet's mineral wealth and how we should view this issue within the framework of sustainability. The book goes into the history of how human society has used minerals, their relationship with the evolution of human civilization, and how we should use these resources in the future. There is a wealth of information in this volume that deals with important minerals like uranium, lithium, rare earths, copper, nickel, zinc, phosphorous, and others. Readers would find the material presented very informative and a valuable basis for discussions on minerals policy."

—Rajendra K. Pachauri, chairman,
UN Intergovernmental Panel on Climate Change;
CEO, The Energy and Resources Institute

UGO BARDI

EXTRACTED

How the Quest for Mineral Wealth Is Plundering the Planet

FOREWORD BY JORGEN RANDERS

A Report to the Club of Rome

Chelsea Green Publishing
White River Junction, Vermont

Originally published in German as *Der geplünderte Planet* in 2013 by oekom verlag GmbH, Waltherstraße 29, 80337 München.

Editor: Joni Praded
Project Manager: Patricia Stone
Copy Editor: Nancy Ringer
Proofreader: Eileen Clawson
Indexer: Shana Milkie
Designer: Melissa Jacobson

Printed in the United States of America
First printing April, 2014
10 9 8 7 6 5 4 3 2 1   14 15 16 17 18

Chelsea Green Publishing is committed to preserving ancient forests and natural resources. We elected to print this title on 30-percent postconsumer recycled paper, processed chlorine-free. As a result, for this printing, we have saved:

18 Trees (40' tall and 6-8" diameter)
9 Million BTUs of Total Energy
1,554 Pounds of Greenhouse Gases
8,431 Gallons of Wastewater
564 Pounds of Solid Waste

Chelsea Green Publishing made this paper choice because we and our printer, Thomson-Shore, Inc., are members of the Green Press Initiative, a nonprofit program dedicated to supporting authors, publishers, and suppliers in their efforts to reduce their use of fiber obtained from endangered forests. For more information, visit: www.greenpressinitiative.org.

Environmental impact estimates were made using the Environmental Defense Paper Calculator. For more information visit: www.papercalculator.org.

Our Commitment to Green Publishing

Chelsea Green sees publishing as a tool for cultural change and ecological stewardship. We strive to align our book manufacturing practices with our editorial mission and to reduce the impact of our business enterprise in the environment. We print our books and catalogs on chlorine-free recycled paper, using vegetable-based inks whenever possible. This book may cost slightly more because it was printed on paper that contains recycled fiber, and we hope you'll agree that it's worth it. Chelsea Green is a member of the Green Press Initiative (www.greenpressinitiative.org), a nonprofit coalition of publishers, manufacturers, and authors working to protect the world's endangered forests and conserve natural resources. *Extracted* was printed on paper supplied by Thomson-Shore that contains at least 30% postconsumer recycled fiber.

Library of Congress Cataloging-in-Publication Data
Bardi, Ugo.
Extracted : how the quest for mineral wealth is plundering the planet : a report to the Club of Rome / Ugo Bardi.
pages cm.
Includes index.
ISBN 978-1-60358-541-5 (pbk.) — ISBN 978-1-60358-542-2 (ebook)
1. Mineral industries—Environmental aspects. 2. Mines and mineral resources. 3. Conservation of natural resources. I. Club of Rome. II. Title.

TD195.M5B368 2014
333.8—dc23

2014000242

Chelsea Green Publishing
85 North Main Street, Suite 120
White River Junction, VT 05001
(802) 295-6300
www.chelseagreen.com

To my son Francesco, the geologist.

A Message from the Club of Rome

Extracted: How the Quest for Mineral Wealth Is Plundering the Planet is a Report to the Club of Rome. It is peer reviewed by the Club of Rome and its expert members to ensure that it is scientifically rigorous and innovative and contributes a new, important element to the debate about humanity's predicaments. Since *The Limits to Growth*, the first Report to the Club of Rome in 1972, 33 publications have received this imprimatur.

The Club of Rome was founded in 1968 as an association of leading independent thinkers from politics, business, and science. It now has 150 individual members; an international center in Winterthur, Switzerland; and national associations in 30 countries. An important element of the national associations' work is to shape national agendas.

Members are unified in their concern for the future of humanity and the planet, and in their goal to address the root causes of the systemic crisis. Their work focuses on the need for a different set of values to change economic theory and practice and safeguard resources; the creation of a more equal society that generates full employment; and the need for governance systems that put people at their center. This holistic approach is needed now more than ever before.

The club pursues its objectives through scientific analysis, communication, networking, advocacy, and cooperation with a wide range of partners. Its main products are books, discussion papers, policy briefs, conferences, webinars, lectures, high-level meetings, and events. Key findings are used to challenge policy makers in the public and private sectors to shift to new ways of thinking and new forms of action.

With *Extracted*, Ugo Bardi presents the current state of knowledge and advances the debate around the issues of depletion and misuse of our planet's natural capital. Since the founding of the Club of Rome in 1968, the question of humanity's growth and resource use in relationship to our planetary boundaries has been central to its work. Two recent Club of Rome reports complement this vital debate: Ernst von Weizsäcker's *Factor Five*, which shows how meaningful action in the coming decades can transform the global economy through an 80 percent improvement in resource productivity, and Gunter Pauli's *The Blue Economy*, which presents business models that can shift society from a state of scarcity to one of abundance by tackling, in new ways, issues that cause environmental and related problems.

Contents

Glimpses

Foreword

More than forty years ago, I was part of an MIT team that set out to understand the potential long-term consequences of various global policies, like those surrounding population growth and economic growth. We wanted to understand what actions could lead to a future where humans lived in balance with nature, and what actions could lead us to overshoot our planet's natural limits, ultimately reducing its carrying capacity. Move a lever in this or that direction, and what would happen? It was my first deep look into how the interplay of physical realities and human behaviors can lead to multiple possible outcomes.

Our model-driven study became known as *The Limits to Growth*, summarized in a book of the same name. Among the many questions it probed was how the depletion of nonrenewable mineral resources would affect the world's economy over a time span of more than a century. Were we likely to "run out" of critical minerals? We found that was unlikely: our models showed that mineral depletion starts affecting the economy long before minerals disappear. Why? Because we would most likely run out of the capital needed to exploit minerals before we ran out of the minerals themselves. Our data suggested that, as a consequence, mineral production would begin to decline within the first few decades of the twenty-first century. In the long run, we would leave significant amounts of mineral resources unexploited underground.

So here we are, at that threshold. We have certainly dug and drilled our way to various environmental problems, but what is the outlook for our mineral resources themselves?

In *Extracted*, Ugo Bardi examines again the phenomenon of depletion. Most books on mineral resources tend to be tedious lists of reserves, lined up as if they were soldiers ready for battle. But Bardi takes a different approach. In the pages ahead, he recounts the whole sweeping story of minerals, beginning with their creation in the giant explosion of supernovas. He shows us how ancient and slow geological processes accumulated them in ores in the earth's crust. And he recounts how humans found this hidden treasure, how it made and changed civilizations, and how in many cases we plundered it with little regard for the consequences to the ecosphere and to ourselves. At a time when discussion of mineral depletion often resorts to black-and-white analyses of what we are running out of, what has peaked, and how we might cope without it, *Extracted* offers a full-bodied analysis that illuminates the real consequences of relentlessly plundering the planet for its mineral riches: an altered landscape, massive pollution issues, potential economic upheaval, and, among other serious results, the unleashing of greenhouse gases by mining and burning fossil fuels.

Forty years of watching environmental, economic, and behavioral trends lead me to believe that, on the fossil-fuel front, the costs of mitigating climate impacts will lead us to stop unearthing coal, oil, and gas well before we run out of them—albeit, not soon enough to prevent serious damage. I suspect similar economic constraints will keep us from exploiting the last of the other critical minerals discussed in this book as well. But that doesn't mean depletion is not a concern.

For instance, Bardi emphasizes that the depletion of fossil fuels is not "solving" the problem of climate change. Rather, at present, it is making it worse—because as easy-to-access sources of oil and gas grow more scarce, the industry has begun to extract from more-polluting sources. As I describe in my book *2052*, it will likely take a few decades before the combination of depletion, economic decline, and population decrease leads to a substantial decrease in greenhouse-gas emissions. In the meantime, the twin problems of depletion and climate change must be faced, understood, and acted upon, or we will badly suffer from both.

In reality, depletion is a long-term phenomenon, a ponderous series of steps that continues for decades and centuries, but it is the unfortunate tendency of the human mind—and especially of the political and corporate mind—to see only the short-term future and make decisions based on short-term gains. Bardi, though, gives us a long-term view, explains how depletion is already playing a significant role in our world, and explores some of the changes we'd need to make, economically and politically, to arrive at a better future than the one we're currently heading toward.

Jorgen Randers
January, 2014

Preface

The saga of mining began tens of thousands of years ago, when our remote ancestors started digging for the stones they used as tools. It was a humble beginning for a revolution that led to the modern mining industry, which today extracts and processes billions of tons of materials every year. This gigantic flow of mineral commodities provides the energy and vital resources needed for the world's industrial economy to continue producing goods and services.

But, as the Earth is plundered of its mineral treasures, fears about "running out" of critical minerals have been voiced more and more frequently. These fears have been often ridiculed as the opinion of Cassandras, from the name of the mythic prophetess who was cursed by the gods to be never believed.

However, we cannot forget that the Earth is a finite planet, as are the veins, the ores, the seams, and the wells from which we are extracting minerals. It is legitimate to ask how long these supplies can last. It is also legitimate to ask how the gradual depletion of mineral ores will affect the economy—even long before we actually "run out" of anything. And, finally, it is even more legitimate to ask how the dispersal of the mined materials, something that we define as "pollution," will affect the Earth's ecosystem. Many of these materials are poisonous for living beings, and many of the chemicals used to extract them are toxic or damage the environment. When it comes to mining fossil hydrocarbons like coal, oil, and gas, the impacts take an even more dangerous turn, as the ultimate end result is the release of carbon dioxide (CO_2), which is irreversibly altering our planet's climate.

Without doubt, mining activities have dramatically reshaped our planet—even our physical landscapes—and fueled an economy bent on endless growth that depends on a seemingly endless supply of raw materials. Everything we use, after all, if not grown, must be mined. But how long can the supply of minerals last? We all know, at some level, that it cannot last forever. We live on a finite planet. Even so, people, industries, and governments that rely on finite resources are often loath to take a true, hard look at just how plentiful or scarce certain resources are, not to mention the consequences of mining or using them. We remain, as a society, reluctant to accept natural limits, particularly when those limits challenge the notion that we can continue on with business as usual.

One of the first studies that attempted to analyze and quantify these issues was *The Limits to Growth*, published in 1972.[1] It was sponsored by the

Club of Rome, a think tank of intellectuals concerned about the world's future, and it was carried out by a group of researchers of the Massachusetts Institute of Technology. Using the best computers of the time, the *Limits* study took into account the interaction of several parameters of the world's economic system and developed scenarios for its possible evolution up to the end of the 21st century. It considered everything from resource availability to population growth and a host of other factors, including the increasing costs of extraction and the increasing costs of fighting the pollution created by industrial processes. The goal was to present whole-picture scenarios—an approach that had not been attempted before and that could map out probable consequences over time of the combined effects of depletion, pollution, and population growth.

The results left little space for optimism: resource depletion and damage resulting from pollution were bound to stop economic growth and generate the irreversible decline of the industrial and agricultural systems at some point in a not-too-remote future. That, in turn, would generate the decline of the human population. The "base case" scenario, the one that used the data that were considered to be the most reliable at the time, showed the industrial and agricultural decline beginning in the first decades of the 21st century, followed by the start of the population decline some decades later. Other scenarios, based on different estimates of the input parameters, generated a later decline but could not avoid its occurrence, even with very optimistic initial assumptions. The study showed that only radical changes in the way the world's economy was run could avoid the decline and stabilize the economic system over the long run. To reach this goal, the authors recommended measures such as putting a limit on industrial growth and the extraction of mineral resources. They also recommended sustainable practices in industry and in agriculture, as well as measures to limit population growth.

It goes without saying that none of these measures was ever put into practice. The story of *The Limits to Growth* is not only about an academic study but also about how difficult it is for our society to plan for the future. The publication of the book generated a hot debate that, in some years, degenerated in all-out smear campaigns aimed at destroying the credibility of the study. Eventually the public became convinced that the *Limits* study had been nothing more than a series of wrong predictions prepared by a group of deluded scientists who had thought that we were soon to run out of everything.

But the public perception of the *Limits* message was wrong; none of the scenarios developed in the *Limits* study predicted that humankind would run out of anything before the end of the 21st century. The scenarios, instead, were based on the obvious concept that progressive depletion could only cause an increase in the costs of production, while the accumulation of waste would cause an increase in the costs of fighting pollution. While proponents of unchecked growth continue to fiercely condemn the results, *The Limits to Growth* and its updates in 1982 and 2004 have been examined and validated by later studies.[2] In fact, various studies have shown that the trajectory of the world's economic parameters has followed the base-case model rather closely.[3] That "base case" scenario estimated that pollution and depletion together would start becoming a stumbling block to economic growth sometime between 2000 and 2020, and that may explain the turmoil in the world's economy that we are seeing nowadays. Like Cassandra's, the authors' warning has rung true.

But that doesn't change the fact that important ground was lost while naysayers considered the study a threat to business as usual. Eventually, and unfortunately, systemic studies on depletion and economy were largely abandoned in the wake of the optimism of the 1990s, when, for a while, most people seemed to believe that the Internet was going to bring us an everlasting era of infinite prosperity.

Today, interest in the theme of resource depletion has renewed.[4] Several studies have concluded that we are, indeed, approaching a point at which the gradual depletion of low-cost mineral resources is becoming a major limitation to economic growth and even to maintaining the present level of economic output. The problem of dwindling mineral resources is all the more crucial because it is arriving in tandem with accelerating ecosystem disruption and rapid growth of the human population. Global temperatures are rising, severe weather events caused by climate change are increasing, and a host of further problems, from ocean acidification to droughts and loss of biodiversity, are before us.

These problems can't just be boiled down to the perils of "running out of something" or of a modest increase in atmospheric temperatures. Instead, they represent a complete transformation of the whole Earth's ecosystem, generated by the human influence on the planet. So, the call to action urged in the 1972 *Limits* study is becoming more and more urgent. We need to face the problems of ecosystem disruption and mineral depletion with better efficiency in all sectors of industry, with the use of renewable resources, and with the development of effective recycling processes to lengthen the life of the remaining resources. Acting effectively against these problems requires a functioning industrial economy that can provide the resources necessary to begin substituting non-carbon-based energy sources for fossil fuels, as well as for mitigation measures (and perhaps geoengineering) against the damage cause by climate change. Only in this way can we face the twin challenges of depletion and climate change.

The pages ahead offer a sweeping look at the history of mining, along with a systemic and scientific look at the current state of mineral depletion and its effects on the economy and the ecosystem. Part 1 examines the great cycle of mining that started tens of thousands of years ago and shows signs of being in the process of winding down. It explores the ancient processes that created minerals, the history of mining, and the rise of mineral empires. Part 2 delves into the marriage of minerals and energy, examines how we model depletion, and probes the dark side of our reliance on continual extraction. Part 3 considers the shape of things to come, investigating strategies for maintaining society's energy and other needs without the supplies of cheap mineral commodities that we have been used to having until now.

Throughout the book, "glimpses" provided by various minerals experts probe the future of certain minerals—detailing what remains, what can be reasonably extracted, what effects supply levels will have on the economy, what can be recovered from material already in use through recycling, and what can be substituted. Many mineral resources are presently marketed in the world's economy. The US Geological Survey lists some 90 of them in a yearly updated assessment. The aim here is not to repeat that listing, but rather to evaluate selected critical minerals—those that carry special importance as energy sources (like fossil fuels and uranium), in infrastructure and manufacturing (like nickel, zinc, and copper, among others), or in high-tech applications (like rare earths and lithium). Other glimpses look at the supplies of minerals that affect food security, as phosphate does. Some of these glimpses take a look at sweeping changes that are taking place right now in the world's economy, and all have a long-range perspective and concentrate on worldwide trends.

The conclusion of this assessment is that we are bumping up against limits on a number of these critical resources—some sooner than others—and that the methods the global mining industry uses to forecast remaining supplies may be entirely inadequate when it comes to determining how many of those supplies can be extracted without unbearable cost—financially, environmentally, and in terms of energy.

PART ONE

HOW IT ALL BEGAN

In ancient times, the underworld was often seen as a place of punishment and suffering. This illustration by Gustave Doré depicts the underworld described in Dante Alighieri's epic poem, *The Divine Comedy*.

1

Gaia's Gift:
The Origin of Minerals

For our ancestors of long ago, the depths of the Earth must have been a source of great fascination. Volcanoes, earthquakes, geysers, hot springs—all were manifestations of the powers residing underground. Clearly, the Earth moved, it quaked, and it spewed out gases and vapors. It must have seemed to be somehow "alive." But what exactly was the source of that power? The lack of suitable tools to dig to any significant depth left our ancestors without clues to the features of the underworld, except for what they could observe by exploring natural caves. Those explorations must have stimulated their imagination. It is no surprise that in the late Paleolithic period caves were used for rituals and for creating those paintings of hunting scenes that we can still admire today.

With the appearance of agricultural civilizations, the underworld became part of the world's mythological pantheons. In those ancient times, many believed that immense powers resided there—like the power of a volcano embodied by the Greek Chimera, a mythical fire-breathing monster.[1] People had to use fantasy to make up for the lack of known fact, and the first written story of a trip to the underworld is a myth that dates back to the third millennium BCE. In it, Inanna, the Sumerian goddess of fertility, visits a dark world of caverns populated by monsters, demons, and unfriendly deities. Such underworld stories are rife with souls of the dead wandering forever in the obscure landscapes of the depths below. In an early Mesopotamian story, the dead dwell beneath the Earth, "eating clay and drinking dust."[2] In the myth of Orpheus, the hero attempts to bring his loved one back from the underworld but fails, a theme repeated in many other myths. Millennia afterward, Dante's *Divine Comedy* (14th century CE) still described the underworld as a place where the souls of the dead resided, forever punished for the sins they committed in life.

Apart from myths, there were already in ancient times practical reasons for being fascinated with the underworld. Even Stone Age people knew very well that rocks were not all the same: some could be used for tools, others for paintings, others for lighting fires, and more. But the variety of rocks that could be found went beyond practical uses. There were spectacular crystals, often translucent and brilliantly colored, that later became known as gemstones. There were shiny chunks that appeared in the sand of riverbeds—nuggets that today are universally recognized as copper, silver, and gold. Eventually it was found that these metals could be worked into different shapes to make tools or elaborate jewelry. And later it was found that some rocks could be transformed into something completely different by heating them at high temperatures. All of these discoveries surely led to questions about the origin of minerals, but in the early history of mining no good answer could be found.

The Birth of a New Science

In time, knowledge about the properties of the underground started accumulating, and the first theories about the origins of minerals were developed. Theophrastus, an ancient Greek, and Pliny the Elder, an ancient Roman, wrote at length about the properties of minerals known during their times but were at a loss when it came to understanding their origins. The main theory in those days was developed by the Greek philosopher Aristotle and was based on the idea that minerals formed when some kind of gas exhalations from the depths of the Earth solidified. According to this view, minerals would grow with time, just as living beings do. So minerals might well re-form in the places where they had been extracted, just as plants would re-grow after having been harvested. The concept of "mineral depletion" as an irreversible process was unknown to the ancients, even though they did note that individual mines tended to run out of the ores they contained.

It wasn't until the Renaissance, when Georg Bauer arrived on the scene, that the origins of minerals were investigated with a scientific approach. Bauer, under the pseudonym Agricola, wrote his *De Re Metallica* (On the Nature of Metals) in 1556. It was a milestone in the science of mineralogy, and it put to rest forever the idea that minerals were living creatures. Bauer's work was expanded upon by the early pioneers of modern geology like Nicolas Steno, Georges-Louis Leclerc de Buffon, William Hutton, and many others.

At the beginning, geologists had to battle a stiff resistance to the concept that the Earth is much older than the Bible says it was. In a way, their task was much more difficult than that of astronomers trying to establish the reality of the heliocentric system. After all, Galileo had to fight only a line in the Book of Genesis that says that the Earth stands still; geologists had to fight the whole book, since it says that the Earth was created over six days some four thousand years ago and has remained static ever since. Some people today still remain wedded to a literal interpretation of the biblical creation story. However, geology has moved forward, and consensus was gradually obtained on the fact that the Earth is billions of years old.

During the past century or so, the revolution in Earth sciences begun by the early pioneers has continued, and a fascinating picture of the Earth's history has unfolded in its wake. Our planet now appears to us as a dynamic entity, almost a living being, where geological and biological forces combine to maintain conditions that support biological life. A big shift in our understanding came in the early 20th century, when Alfred Wegener introduced the concept of "continental drift" (later renamed "plate tectonics"), a fundamental element of the Earth's system dynamics.[3]

In time, the purely geological view of the Earth system merged with the idea that biological organisms interact with their inorganic surroundings to create a global system, called Gaia, that is dominated by self-regulating feedbacks and is constantly changing and adapting to maintain conditions that make life on Earth possible. The concept of Gaia has been gradually making inroads in established scientific thought, although it remains somewhat controversial.[4] One problem is the difficulty of defining exactly what is meant by "Gaia," and the concept has evolved considerably since it was first proposed. In particular, Gaia theory cannot be understood today without taking into account the stabilizing effect of geological cycles, and some recent criticism misses this important point.[5]

In any case, the fact that the name Gaia comes from the ancient Latin Earth divinity has generated plenty of confusion. Some people have cried blasphemy.[6] Others assume that those endorsing the Gaia theory are something akin to a divinity cult, complete with festivals and rituals.[7] Of course, that never was the intention of the term. Gaia, or the Earth system, is not a deity or even a sentient being, and "she" has no interest in the survival or well-being of human beings or of any living creatures in general. So, it is rather useless to worship Gaia as a goddess or even to say that Gaia somehow "optimizes" the environment for living beings. But it makes plenty of sense to note the existence of important stabilizing feedbacks in the Earth's systems. Using the term "Gaia" is a convenient way to label this set of feedbacks. In this sense, Gaia shares some, though not all, of the characteristics of living creatures.

One of the consequences of Gaia's active cycles is the formation of mineral ores and deposits, entities that we could call "Gaia's gift," as they are the result of planetary forces that have been active for billions of years. But in order to understand the origin of mineral deposits we must start from the beginning of a very long story.

A Planet Is Born

Some 4.6 billion years ago the solar system formed, resulting from the condensation of a cloud of debris left in space by the explosion of ancient supernovas. Our sun is a second-generation star, which means that the mix of gases that created it—and the planets in the solar system—contained a certain amount of heavy elements that had formed inside the fiery heat of the supernova explosions. It was the presence of these heavy elements that generated the rocky planets of the solar system, including Earth.

The condensation of Earth to form a solid planet marks the start of the geological period that we call Hadean (named after Hades, the ancient Greek underworld). As the planet formed and gained mass, gravitational energy was released and its temperature increased. Eventually the planet became so hot that it melted. In this phase the heavy metals, mainly iron and nickel, sank to the center, taking with them the elements that would easily dissolve in molten iron. Some light elements, mainly silicon, aluminum, and oxygen, formed compounds not easily dissolved in the core and were left mostly in the outer shell in the form of oxides. This event is sometimes called the "iron catastrophe."[8] Afterward, the surface of the planet cooled relatively rapidly, and it appears that by about 4.2 billion years ago Earth had a solid surface and an inner structure not unlike the present one: a hot metal core and a relatively cold silicate outer shell, or mantle.

During the last phase of the Hadean eon, around 4 billion years ago, the data indicate the occurrence of a period of intense asteroidal bombardment that may have partly restored the concentration of heavy metals at the surface, making most of today's mining possible.[9] The bombardment may also have brought to Earth the mass of water that still forms our oceans.[10] Life may have originated during this period, perhaps at volcanic undersea vents where living creatures could exploit the chemical energy contained in the compounds, mainly sulfides, generated by the heat of the mantle.[11]

This ancient world had some similarities to our own but was also very different. It was covered almost completely with water, and volcanic activity must have been rampant. The small patches of land surface, if there were any, showed no trace of macroscopic life forms, and the atmosphere contained no oxygen, or just traces of it. The moon is believed to have been much closer to Earth than it is today, and this proximity must have raised gigantic tides that periodically swept the edges—or perhaps the whole—of the land masses.

The presence of liquid water during the Hadean eon raises a problem called the "paradox of the faint young sun."[12] Our understanding of the life of stars tells us that the sun of that ancient time must have been about 30 percent colder than it is today. From this, we can calculate that Earth's temperatures should have been too low to maintain liquid water on its surface. Earth should have been a frozen ball of ice—like Europa, the moon of Jupiter, is today. There are various possible explanations for Earth's unexpected warmth: it may be related to the presence of greenhouse gases in the atmosphere or to special characteristics of the early sun. At present, the most likely hypothesis seems to be that it was mainly due to the large fraction of young Earth's surface that was occupied by oceans. Since water absorbs sunlight better than solid ground, the oceans could have absorbed enough heat to maintain relatively high temperatures.[13]

The Hadean was followed by the Archean eon, which started 3.8 billion years ago and was a much quieter period in terms of planetary changes. Nevertheless, the heat flow from Earth's nucleus was still two to three times greater than it is today and volcanic activity must have been frequent and intense. The Archean saw the rise of the modern continental land masses. This process involved the accretion of low-density, silica-rich materials that, being lighter than the average oceanic crust, tended to "float" over it. The silica-rich rocks formed granitic solid bodies, which were the origin of the present continents. These proto-continents are generally believed to have been much smaller than the present ones, although one hypothesis suggests that continents of about the same size as the present ones formed very early in Earth's history.[14] In any case, the Archean oceans are also likely to have contained more water than their present-day counterparts, perhaps as much as three times more.[15] The Earth of the Archean eon, therefore, was a planet mainly covered with oceans.

Over the ages, much of this Archean water has been lost. The ocean basins are, in a way, "leaking" to the underlying mantle by a process that involves the formation of silicate hydrate (that is, water-containing) compounds that are pushed into the mantle at the edges of continents by the continuous movement of plate tectonics. Another mechanism that causes the loss of water from Earth's surface ecosystem is photodissociation, the breakdown of water molecules under the effect of ultraviolet light generated by the sun. This process generates hydrogen and oxygen; the former may escape to outer space, and therefore water cannot be re-formed again. Over geological time scales, these phenomena have gradually reduced the amount of water at Earth's surface and caused the gradual emergence of the land masses that we know today.

During the Archean, radiation from the sun was still considerably less than it is nowadays, but the ocean's low ability to reflect heat and, possibly, the presence of high concentrations of greenhouse gases (mainly CO_2) in the atmosphere kept temperatures high enough to maintain liquid water at the surface. The climate of the Archean is traditionally believed to have been considerably warmer than the present one, but recent studies indicate a more temperate climate and perhaps the occurrence of ice ages during the last part of the period.[16]

Life during the Archean existed in the form of simple single-cell organisms in the oceans. Life's metabolism was already supported by photosynthesis and the by-product of this activity was oxygen, a gas that would have been poisonous for the organisms of the time. However, the amount of oxygen in the atmosphere remained low. It is likely that oxygen was removed from the atmosphere as soon as it formed by reaction with minerals, such as the iron ions dissolved in the oceans. The result was the formation of solid iron oxides that then sedimented at the bottom of the oceans. It is from these ancient layers that we are extracting most of the iron produced today.

The Archean lasted until 2.4 billion years ago, when the so-called "great oxygenation event"[17] ushered in the Proterozoic eon, introducing major changes in Earth's atmospheric composition and the way its ecosystems functioned. It seems that the great oxygenation event came about largely due to saturation of the iron sinks that had been removing the oxygen produced by photosynthesis during the Archean eon. However, the phenomenon is probably more complex and not yet fully understood. In any case, microorganisms learned how to exploit the growing amounts of oxygen to boost their metabolism. A planet once populated by anaerobic (that is, not needing oxygen) life forms suddenly became oxygen fueled, and life exploded—although still in the form of single-celled organisms living mainly in the oceans.

Around 540 million years ago the Proterozoic eon came to a close, with a new burst of life at the start of the Phanerozoic eon, otherwise known as the age of visible life. As the concentration of oxygen grew in the atmosphere, fish and other marine life began to appear in the oceans, and amphibians and plants on land. In time, the continents were completely colonized by plants and animals.

The Phanerozoic age lasted for more than 500 million years and is still ongoing. It saw several dramatic climatic changes, ranging from ice ages to balmy periods when Earth was a veritable hot greenhouse sometimes described as "hothouse Earth."[18] It saw gigantic volcanic eruptions and massive asteroidal impacts. Life survived and rebounded from these catastrophic events in a series of changes that are often described as a continuous progress toward higher forms of life. It is also true, however, that biological productivity on the planet may have peaked long ago, during the first phase of the Phanerozoic, known as the Paleozoic, and gradually declined afterward. Optimal conditions for life may have occurred during that period as the result of the balance of solar irradiation and carbon dioxide concentration.[19]

The last phase of the Phanerozoic is known as the Holocene—the past 12,000 years of life of the planet, which have seen a relatively stable climate and the development of human civilization. The last period of the Holocene is often referred to as the Anthropocene, though this term is not yet officially recognized. In any case, the Anthropocene is defined as the period in which the effects of human activities—including agriculture, mining, increasing population, and pollution—on Earth's ecosystem have become noticeable and even preeminent, coming together to initiate the development of a new ecosystem whose characteristics have yet to be fully revealed and may be not at all positive from the viewpoint of human beings. In order to understand how these effects are acting and how Earth is changing, we need to understand how the ecosystem works—the inner mechanisms of Gaia as a living planet.

Gaia: The Living Planet

Today, the inner structure of our planet has not changed much from the early Archean times, at least in qualitative terms. The core temperature has cooled, but it remains high enough to maintain a metallic hot nucleus, partly molten and partly solid, because of the tremendous pressure exerted upon it by the weight of the earth above it. The present temperature of the inner core is believed to be around 6,000 degrees Celsius. Some of this heat lingers from that created in the formation of the original proto-planet, but most is created by the decay of radioactive isotopes such as uranium and thorium.[20] The metallic core is surrounded by the thick mantle shell, formed mainly of silicates, minerals that combine oxygen and silicon. The mantle temperature is about 4,000 degrees Celsius closest to the core, and between 500 and 900 degrees closest to the crust.

The heat flow from the core of Earth is fundamental for shaping the world as we see it today. This flow is small, amounting to only about one-tenth of a watt per square meter; but measured across the whole of Earth, that energy amounts to 44 terawatts, significantly more than the energy generated by human beings today, mainly by fossil fuels. This heat is large enough to generate a series of geological phenomena that keep Earth "alive." Without this internal heat, Earth would be a dead planet, just like the moon and Mars are.

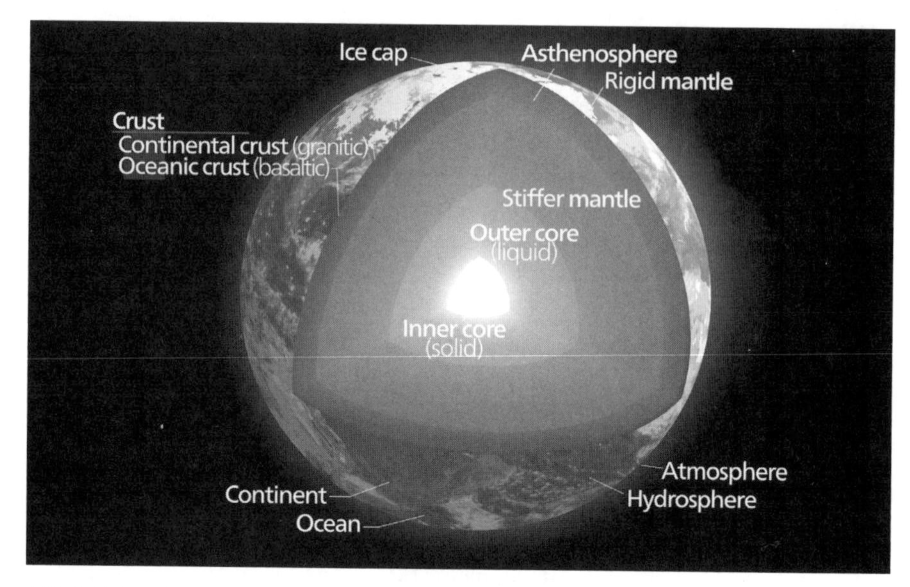

FIGURE 1.1. The inner structure of Earth as it is today.

Heat flowing from the core creates convective movements in the viscous, semi-molten mantle. These movements are the source of most geological activity, from volcanoes to earthquakes, experienced at the surface of the planet. Upward convection flows generate ridges at the bottom of oceans, where mantle material is continuously pushed to the surface of the oceanic crust. There, it cools down, is pushed aside by the arrival of new material from below, and moves away on a de facto conveyor belt that starts at the ridge and arrives at the edge of continents, where it is pushed back inside the mantle by a process called subduction. The whole trip over the oceanic crust may last tens of millions of years. The continents are continuously pushed around by the convective movements of the mantle at very slow speeds: just a few centimeters per year, slower than the growth of human hair or fingernails. But they do move, and over billions of years continental masses have performed a complex dance that has seen them separate and reunite in a series of gigantic fractures and clashes. The continents that we see today drifted apart from an ancient supercontinent, called Pangea, that started breaking down some 170 million years ago.

When one continent bumps into another, the collision usually lasts for millions of years and involves enormous amounts of energy. The process causes the crust to corrugate as large amounts of material are pushed against each other and pile up, forming what we see as mountain ranges. The Himalayas, for instance, are the result of the collision of the Indian plate against the Asian plate—a process that started about 50 million years ago and is still ongoing. The European Alps erupted from the northward movement of the African plate, which will eventually destroy the Mediterranean Sea. This is why chunks of ocean floor and marine fossils can be found in mountains. These fossils puzzled ancient geologists, who had no other solution than to attribute them to the biblical Great Flood.

An enormous amount of energy is associated with the movement of the oceanic conveyor belt, and this energy builds up pressure against the rigid edges of the continents. When released, this energy generates volcanoes, earthquakes, and the associated tsunamis. The rock pushed into the mantle at subduction zones contains water embedded in silicate hydrate rocks. At the high temperatures of the mantle, these silicates partly decompose, releasing water in the form of a supercritically hot fluid—so hot that it is neither liquid nor gas. This fluid lubricates the movement of the tectonic plates. Without it, the slow movement of continents would literally grind to a halt.

The subduction of water into the hot mantle also builds up pressure that must be released in some way. That water returns to the surface in the form of volcanoes, geysers, hot springs, and other explosive eruptions. Because of this series of processes, water is continuously cycled from the atmosphere to the mantle, and then back from the mantle to the atmosphere.

The cycles generated by plate tectonics are fundamental for the maintenance of the biosphere. Liquid water is needed in order to have living beings and, to have that, planetary temperatures must be maintained within a relatively narrow range. Those temperatures are regulated not by Earth's core but by the sun, and they are strongly affected by the greenhouse effect; that is, by the capability of some atmospheric gases to trap heat emitted by Earth's surface. Without these greenhouse gases—comprising mainly water vapor, but also carbon dioxide and methane—Earth's temperature would be too low to maintain liquid water at the surface. Variations in the concentrations of these gases affect Earth's temperature, and the study of these variations is an extremely rich field that tells us much about the history of the planet.

The most important cycle generated by plate tectonics, though, is the geological carbon cycle, also called the "silicate weathering cycle"—and not to be confused with the biological carbon cycle, which is related to photosynthesis and respiration. "Weathering" is a general term indicating the breakdown of rock under the effect of atmospheric agents and the geological carbon cycle is a complex process that starts with the weathering of common silicate rock in the crust. Carbon dioxide (CO_2), which is slightly acidic, reacts with silicates to

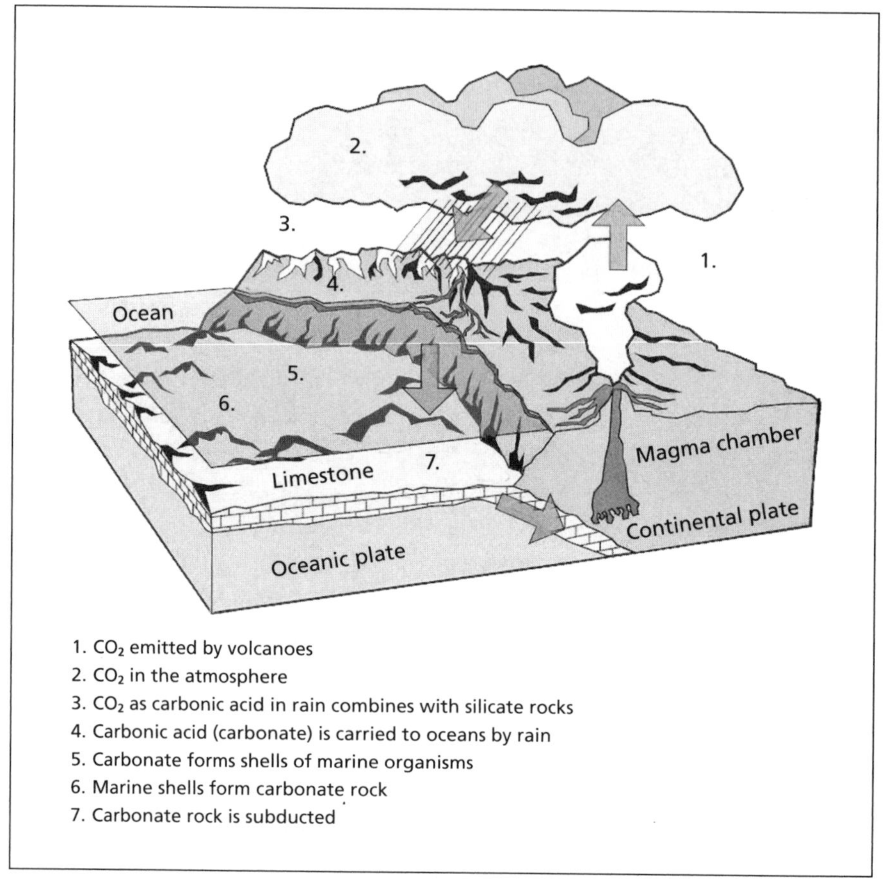

FIGURE 1.2. The geological carbon cycle.

form carbonate rock. The reaction is very slow by human standards, but not so by geological ones, and it gradually consumes atmospheric CO_2. Carbonates are also slightly soluble in water as ions and tend to be transported to the seas and the oceans by rain. There, the carbonate ions may reform as solid carbonates in the shells of marine organisms, which eventually sediment at the bottom of the ocean. Over geological times, the ocean's conveyor belt transports these carbonates to subduction zones, where they are pushed down, inside the mantle. There, at great depths, the high temperatures of the mantle decompose the carbonates, releasing CO_2 that will return to the atmosphere as the result of volcanic activity.

This geological carbon cycle is believed to be the fundamental mechanism for maintaining sufficient carbon dioxide in the atmosphere for plant photosynthesis, without which life on Earth would disappear.[21] Without the effect of volcanoes, all the carbon dioxide in the atmosphere would disappear in a few million years, at most, consumed by the reaction with silicates. But the CO_2 in the atmosphere is continuously renewed, and, since it is a greenhouse gas, the geological carbon cycle regulates temperatures, too. The speed of the cycle depends on surface temperatures. When Earth cools down, volcanic emissions predominate over the removal by silicate weathering and the CO_2 concentration increases, generating a warming effect. The opposite takes place when Earth warms up. So the cycle operates as the true "knob of the thermostat" that has kept Earth's temperature within the limits necessary to maintain liquid water on the surface for billions of years, despite the gradual increase in solar irradiation over the past geological eras.

As can be seen in figure 1.3, the thermostat is not perfect and it can't prevent strong temperature oscillations, but, on the whole, it has prevented Earth's temperature from increasing as the effect of the increasing solar irradiation over these long geological times. There are other factors that may affect Earth's temperature over long time spans. In particular, the gradual sedimentation of organic carbon in the form of compounds such as coal, petroleum, and gas (and their precursor, called kerogen) has removed large amounts of carbon from the atmosphere, with an overall cooling effect that has contrasted the rise in solar irradiation. In relatively recent times (geologically speaking)—that is, during the past 15 million years or so—Earth saw a low-temperature phase characterized by a series of ice ages. It is believed that the main factor driving these ice ages was the rise of the Himalayas, a process that reduced the concentration of CO_2 in the atmosphere through increased reaction with silicates and cooled the planet.

On the other hand, large and long-lasting eruptions (called "large igneous provinces," or LIPs) in remote ages emitted large amounts of CO_2, raising planetary temperatures to levels that would have made it hard for life to cope. LIPs are thought to have caused several major mass extinctions during the Phanerozoic eon.[22] The correlation seems to be very strong, although an alternative hypothesis is that the extinctions were caused by asteroidal impacts, an idea that originated from the discovery of a massive impact that took place around the same time that dinosaurs disappeared.[23] The impact theory has greatly impressed scientists and the public alike, but since that remarkable discovery, no other comparable impact that could be associated with other massive extinctions has been discovered. Debate over the actual cause of extinctions carries on, and it may be that both asteroidal impact and volcanic eruptions were at play in the case of the demise of the dinosaurs.[24] In most

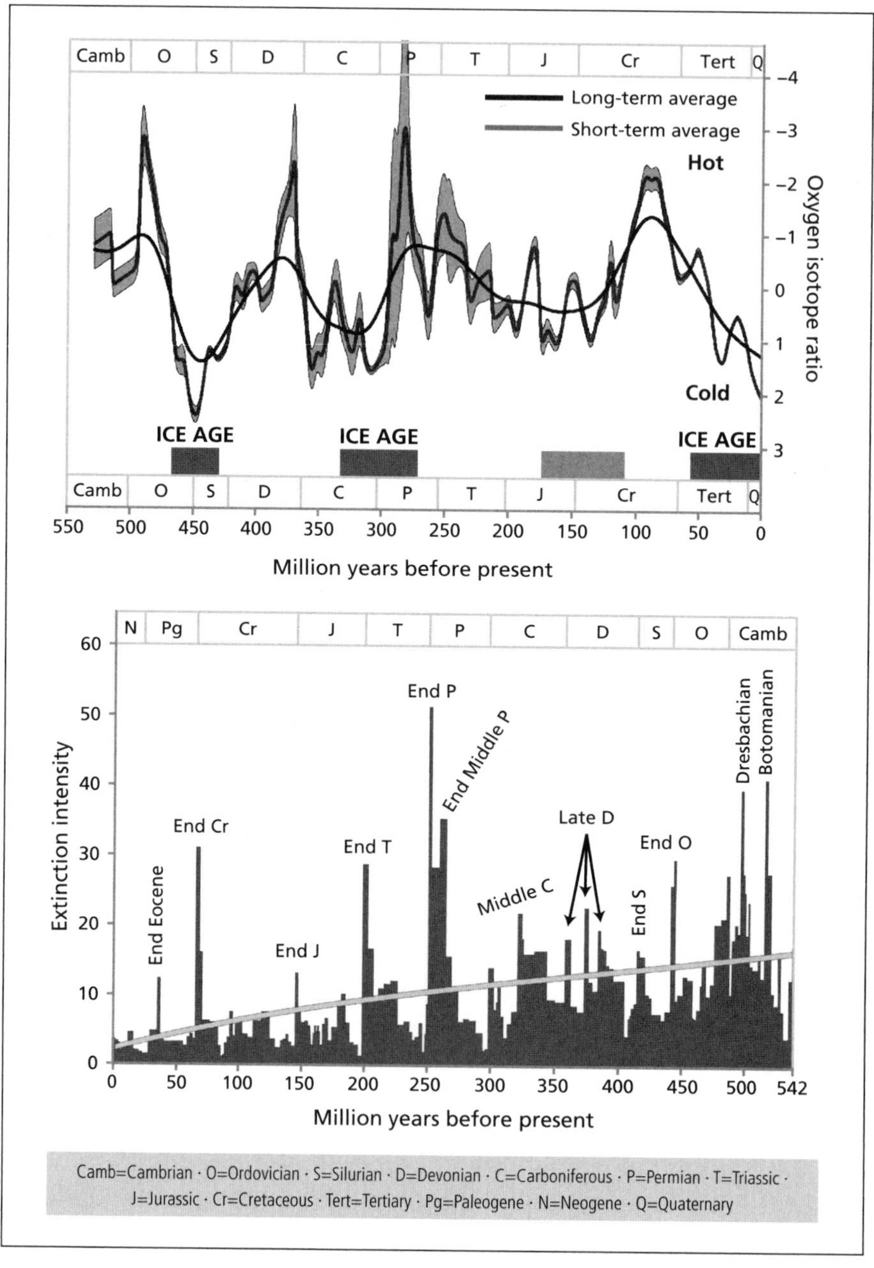

FIGURE 1.3. *Top*, temperatures on Earth during the Phanerozoic age. These temperatures do not show a detectable growing trend, on average, despite the increase in solar irradiation over that period, which should have raised Earth's temperatures. This fact is one of the main proofs of the Gaia concept.

FIGURE 1.4. *Bottom*, extinction intensity on planet Earth during the Phanerozoic eon.

other cases, however, the rapid rise in CO_2 that followed volcanic eruptions was probably the most important element causing these massive extinctions, triggering a series of secondary events that were the actual cause of the extinctions. Life can adapt to changing conditions, but not as fast as greenhouse gas levels can rise and generate global warming, causing the ocean to acidify and lose oxygen, while the rising temperatures spur bacterial activity that emits poisonous hydrogen sulfide. Nevertheless, these spectacular warming events were always followed by a return to less extreme surface temperatures—the effect of the planetary thermostat created by the geological carbon cycle.

CO_2 is not the only known planetary thermostat. There are other greenhouse gases, and there are other factors affecting temperature that are not related to the greenhouse effect—clouds and vegetation cover, for instance. Ice cover, too, can generate climatic effects, reflecting heat, cooling the planet, and thus generating more ice, as it did in those ages when ice completely covered the whole planet surface (called "snowball Earth" phases) for tens or perhaps hundreds of millions of years. But ice doesn't stop volcanoes from pumping CO_2 into the atmosphere and, as a consequence, heating it up so much that the Earth returns to "normal" conditions in a geologically short time, with a dramatically rapid disappearance of the ice cover.

Ores: Gaia's Gift

All that we've discussed up to now is relevant to the origin of mineral deposits. *Deposits* are defined as areas where chemical species that exist in the Earth's crust can be found in greater-than-average concentrations—sometimes several orders of magnitude greater. Those deposits that are concentrated enough to be profitably mined are normally called "ores." (Different terms are used for specific kinds of minerals, such as "wells" for crude oil and "seams" for coal.)

The geological water cycle generates what is perhaps the most important source of mineral deposits on Earth. The supercritical water generated at subduction zones is extremely reactive and dissolves several kinds of metal ions, including those of noble metals such as gold and silver that won't dissolve in water at ordinary temperatures and pressures. This superhot water, laden with dissolved minerals, tends to be pushed to the surface of the crust above, and when it arrives there it is released through volcanoes and hot springs. It then cools, releasing the ions it carries, creating many kinds of high-grade mineral deposits.

This kind of hydrothermal ore formation laid the foundation for human mining. It gave us the noble metal deposits (like gold and silver) and a variety of sulfides (like copper) that spurred human metallurgy. In general, such hydrothermal processes occur only at specific areas; ores, for example, can be found where subduction has occurred in the remote past. This is why the Mediterranean area used to be rich in native copper and gold: it lies at the boundary of the African and European plates, home to extensive subduction and volcanic phenomena. It is also why some of the many thousands who flocked to California in the gold rush of 1849 found what they were looking for: hydrothermal processes had formed gold there when central California was part of an ancient continental edge.

Hydrothermal processes are not alone in creating mineral deposits. A complete description of these mechanisms is the stuff that makes geology textbooks thick, but there are a few others worth mentioning. For a start, hot magma (that is, molten rock) can generate mineral deposits without the need for supercritical water. In this case the mechanism involves dissolving metal elements in molten rock, yielding, for instance, iron, platinum, nickel, chromium, vanadium, and other ores. Diamonds arrive in the crust by entirely differently processes. Carbon compounds that form only at very high temperatures and pressures, in the absence of oxygen, diamonds get their start at great depths inside the mantle. They are transported to the Earth's surface through a rare kind of volcanic pipe in a rock called "kimberlite," from which diamonds can be extracted. From the isotopic composition of diamonds, we know that some of them originated from inorganic carbon present in the early Earth, while others were formed by the condensation of organic carbon that was pushed into the mantle by the subduction process. The latter kind of diamonds are fossils, formed from what was once part of living beings. In any case, all diamonds are billions of years old, and it may be that the processes that formed them are no longer operating because the mantle is cooler than it used to be.

A large variety of lower-temperature processes occurring at the Earth's surface can also form deposits and ores. Perhaps the most important one is the sedimentation of iron in the form of "banded iron,"[25] which contains variable amounts of magnetite and hematite, alternating with bands of sedimentary deposits in the forms known as "shale" and "chert." This kind of deposit is very ancient, created when iron ions that dissolved in ancient oceans combined with the oxygen generated by photosynthesis in blue-green algae. These bands largely ceased to be formed after the great oxygenation event that took place some 2.4 billion years ago, although they reappeared briefly (geologically speaking) in later periods. Ore can also form under low temperatures when a body of water evaporates, leaving evaporates—like ordinary salt—on the ground.

In fact, the number and variety of mineral compounds that we classify as ores and deposits is large enough to be bewildering. But they all have one thing in common: they need energy to form. Deposit formation is thermodynamically uphill; that is, it goes against the trend prescribed by the second law of thermodynamics, congregating rather than dispersing. Without an energy gradient, different chemical species would tend to reach a state of maximum entropy and become well mixed in the crust in forms that would be very hard to mine at a profit. Ores and deposits exist only because the Earth is "alive" and it can provide the energy needed for them to form. In a sense, we could say that ores are "Gaia's gift."

As we've seen, there are two kinds of energy sources that create ores. One, geothermal energy, derives from the Earth. The other derives from the sun, which keeps the geological water cycle ongoing, which dissolves minerals in the form of ions and concentrates them again when the water erupts from the crust (via volcanic activity, for example) and cools. Often both kinds of energy are involved. But there is a third factor that plays an important role as well: the effect of the biosphere. The working of bacteria and other life forms often affects the solubility of metal ions and may greatly speed up the inorganic processes of ore formation. But biology plays the greatest role for human mining by burying carbon, an aspect of the carbon cycle that humans have exploited for their own energy needs. "The Age of Oil" describes the saga that has ensued.

Fossil hydrocarbons and coal formed almost exclusively from decaying organisms. The fate of dead organisms is, normally, to be oxidized by metabolic processes that break down the components of living tissues into water and carbon dioxide, which are then dispersed into the biosphere to be recycled and form new organisms. However, the process is not always complete, especially when oxygen is not present in sufficient amounts. Various stages of degradation and different environments where degradation occurs may lead to different compounds.

Coal, the first fossil fuel used by humans, formed in large amounts during the Carboniferous period, starting about 360 million years ago and lasting for some 60 million years, as the result of the decay and burial of forests. The organic material, mainly lignin, was transformed into peat by degradation in environments poor in oxygen and then, gradually, into the coal deposits we still exploit today. The formation of such large amounts of coal never again occurred in Earth's history, possibly because during the Carboniferous there didn't exist microorganisms able to degrade lignin, which stiffens the cell walls of plants.[26] Today, such organisms rapidly demolish the wood of

The Age of Oil

Colin J. Campbell

Oil and gas were formed in the geological past under well-understood special conditions. So it follows that they are finite resources subject to depletion. It is a simple concept to grasp. As every beer drinker knows, the glass starts full and ends empty. The quicker you drink it, the sooner it is gone. The same applies to oil: for every gallon used, one less remains.

In just a handful of human generations, we have witnessed the birth and, now, the impending death of the age of oil. The first half of this era saw the rapid expansion of oil-based energy, which fueled the growth of industry, transport, trade, and agriculture, allowing the human population to expand sixfold in parallel. But the second half, which now dawns, will likely be marked by a corresponding decline with far-reaching consequences.

It did not take long for the pioneering oil explorers to learn that the discovery of an oil field depended on finding a place where four geological elements came together:

- **Age-old organic matter:** Much of the world's oil comes from just two epochs of global warming, 90 and 150 million years ago, when algae and other organic material proliferated. The remains were preserved in the stagnant depths of lakes and seas in rifts that formed where continents moved apart on the back of deep-seated convection currents in the Earth's crust. The rifts themselves were progressively filled with sediment washed in from the adjoining continents, and when the organic material had been buried to a depth of about 2,000 meters, it became heated enough to be converted to oil. Natural gas was similarly produced from carbonaceous material and also from oil that was overheated by excessive burial.
- **A reservoir:** Once formed, the oil and gas tended to migrate upward to collect in rock that was porous and permeable, like sandstone and limestone. In earlier years it was normal to recover

about 30 percent of the oil in a reservoir, but various sophisticated methods of enhancing recovery have been progressively applied.

- **A trap:** In some places the oil flowed to the surface, where it degraded, with the great tar sands of Canada being a well-known example. But in other cases it was trapped at the top of dome-like geological structures, known as anticlines, or against faults.
- **A seal:** Finally, the reservoir in a trap had to be covered by a seal, principally of clay or salt, to prevent the oil and gas from escaping.

In the early days geologists with no more than a hammer, hand lens, and notebook mapped the outcropping rocks, successfully finding the most promising oil provinces. The world's largest oil province, around the Persian Gulf, was found in 1908 by a well in the foothills of the Zagros Mountains of Iran.

Later there came ever more sophisticated geophysical techniques. An explosive charge was fired, and recorders measured the time it took for the echoes to return from rock surfaces far underground, allowing them to be mapped in detail. Progress in geochemistry also made it possible to test source rocks to identify potential reserves. When the prime prospects of the accessible onshore areas were depleted, the industry turned its eyes offshore, developing ever more sophisticated technology to do so—though only a few offshore areas have the right geology to contain oil or gas.

Once a promising prospect was identified, a rig was brought in to drill what's called a new-field wildcat—a test well on unproven ground. If it confirmed a discovery, the next step was to estimate the oil reservoir's size in order to plan the number of development wells needed to optimize commercial recovery. Pipelines and offshore platforms also had to be planned where necessary. As prices rose, ever smaller fields became viable.

The peak of discovery for so-called regular conventional oil was passed in the 1960s, and extrapolating the long downward trend gives an indication of what remains to be found in the future. Regular conventional has provided most of our oil so far and will dominate all supply far into the future. However, in 1981 we started using more than was found in new fields, and the gap is widening.

How Much Oil Have We Used?

Information on past oil production by country is relatively sound, although war loss has not been reported at all. For example, as much as 2 Gb (billion barrels) went up in smoke in Kuwait in the Gulf War, and that loss should be treated as production in the sense that it depleted the reserves.

Reserve reporting is much less reliable and has been subject to two major distortions. First, in the past the major oil companies found it expedient to report the minimum reserves needed for financial purposes, which delivered an attractive, if somewhat misleading, image of steady growth to the stock market. Those days are, however, now substantially over, because the giant fields, which offer the main scope for underreporting, have matured. The major companies have since found it easier to secure reserves by acquiring existing fields rather than exploring for new ones—leading the largest players, once dubbed the Seven Sisters, to dwindle down to four by merger.

Second, when oil prices fell due to lowered demand in the 1980s, OPEC quotas came under pressure and some nations exaggerated their reserves in an effort to increase the amount of oil they were allowed to produce. In 1985 Kuwait increased its reported reserves from 64 to 90 Gb, although nothing particular had changed in its oil fields. A small, possibly genuine increase to 92 Gb in 1987 proved too much for the other OPEC nations, which promptly announced their own massive increases. Abu Dhabi matched Kuwait exactly (up from 31 Gb), Iran went one better at 93 Gb (up from 49 Gb), and Iraq surpassed both at a rounded 100 Gb (up from 47 Gb). Saudi Arabia could not match Kuwait because it was already reporting more, but in 1990 it held its own by announcing a massive increase of nearly 200 billion barrels . Venezuela for its part jumped from 25 Gb to 56 Gb by including in its reserve figures nonconventional heavy oils that had not qualified for OPEC quotas previously.

A critical element in determining the status of depletion is to identify the different categories of oil and gas, each having its own cost and depletion characteristics. They are broadly described as conventional or nonconventional—although there is no standard boundary. However, "regular conventional" oil and gas can be defined as excluding the following:

- **Heavy oils:** Oils heavier than 17.5° API (a measure of density) and bitumen.
- **Oil shale and shale oil:** Oil shale is immature source rock from which oil can be extracted using heat. Shale oil (also termed "tight oil") is oil that can be produced by artificially fracturing reservoirs lacking adequate natural porosity and permeability.
- **Deepwater oil and gas:** Oil and gas lying in waters deeper than 500 meters.
- **Polar oil and gas:** Oil and gas from the relatively unexplored polar domain (which has certain geological conditions that make it gas-prone); their extraction is subject to high costs.
- **Natural gas liquids:** These liquids are extracted from natural gas in industrial processing plants. (Also known as "natural gas condensate," it is a liquid that naturally condenses from gas and may be conveniently listed with crude oil.)
- **Nonconventional gases:** These include coal-bed methane, hydrates, and shale gas.

Oil production and depletion is a large and complex subject, but the evidence suggests that we are about halfway through the age of oil, as figure 1.5, summarizing some of the factors to be taken into account when evaluating the resource base, shows.

The history of crude oil production is closely related to the great changes undergone by human society. It all started with the advent of settled agriculture about 12,000 years ago. Stone Age man had used flints before people turned to bronze, iron, and steel for better tools and weapons. Minerals and coal were dug from surface pits, which were then deepened into proper mines. The necessity of draining the mines led to a remarkable technological development: the hand pump gave way to the steam pump, which evolved into the steam engine. The steam engine in turn experienced a radical development when a way was found to inject fuel directly into the cylinder, yielding the so-called internal combustion engine, which was much more efficient. For fuel it at first relied on benzene distilled from coal but eventually turned to petroleum refined from crude oil. The first automobiles took to the roads around 1880, and the first tractor plowed its first furrow in 1907. The oil industry grew as demand for fuel increased.

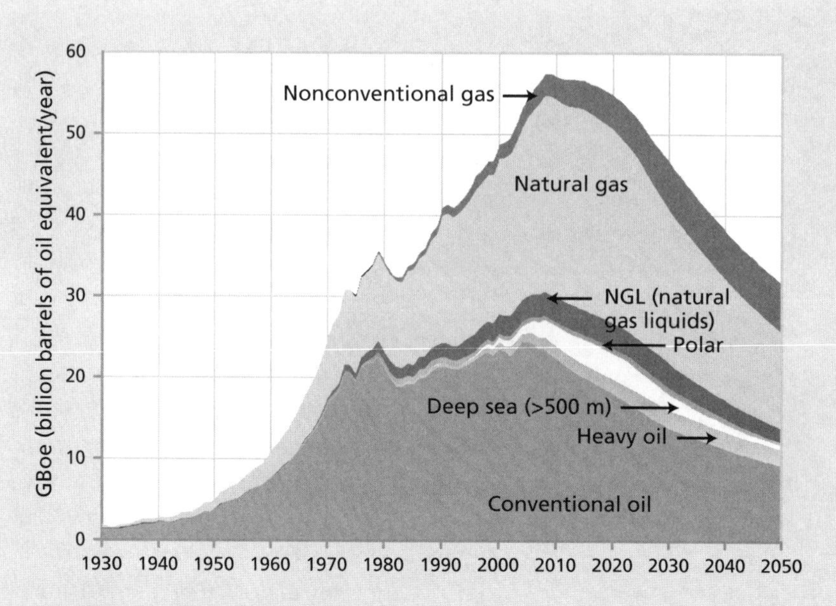

FIGURE 1.5. Oil and gas production profiles.

The growth of industry and population led countries to compete more strongly for trade and expansion of their dominion. At the same time industrial workers began to press for a greater share of the growing wealth. The pressures led to two world wars of unparalleled severity, with access to oil becoming an important issue.

These wars were followed by the so-called Cold War, when the United States vied with its former ally, the Soviet Union, for economic hegemony, progressively adopting the principles of globalism, in which the resources of any country belong to the highest bidder. The US empire was different from earlier ones insofar as it was strictly financial and commercial, having no direct administrative responsibility for territories. It became increasingly dependent on oil imports after its own production peaked in 1970, and it began to take a particular interest in the Middle East. The fall of the shah of Iran in 1979 gave rise to tensions in that region, and the United States supported Iraq, which was engaged in a long border dispute with Iran over oil-rich territory. But that alliance ended when Iraq invaded Kuwait in 1990, to be duly repulsed by a US army. The Middle East remains an area of conflict to this date, its resources vital for the survival of the industrial system of the world.

The Future of Oil

We are far from running out of oil, but the peak of regular conventional oil production was passed in 2005, and the peak for other oil categories will follow shortly (if it has not already been passed). A debate rages over the precise date of the peak but misses the point when what matters is the long decline on the other side of it.

Logic suggests that the economic expansion of the first half of the age of oil will be matched by a corresponding contraction during the second half, given the central place of oil-based energy in the modern world. The peak of regular conventional led to a rise in prices. Shrewd speculators bought contracts on the futures market, whose volume exceeded actual production by factors of ten to thirty, and the industry built storage, watching its oil stocks appreciate in value at little cost. But by 2008, when prices surged to almost $150 a barrel, the traders spotted the limit and started selling short, correctly anticipating that the high prices would trigger recession and cut demand. Prices fell back to 2005 levels before moving up to over $100, a level that seems to have become standard today.

The recession that followed may have been triggered by the oil price shock, and it had devastating consequences. The difficult economic conditions gave rise to riots and revolutions around the world as disillusioned people blamed their governments for what in fact was imposed by nature. And those difficult conditions may continue and evolve. As producing countries come to perceive the depletion of their oil and gas reserves, they will likely restrict exports to preserve as much as possible for their own use, which makes eminent national sense though it offends the principles of globalism. Argentina has already banned exports, and King Abdullah of Saudi Arabia has said that he wishes to leave as much wealth as possible in the ground for his grandsons. The Middle East has become heavily dependent on oil revenue, but its production will have virtually ceased by the end of the present century, underlining the serious tensions that region faces.

A seminal work, *The Limits to Growth*, warned of the unfolding situation of imminent depletion in 1972.[27] Although that study didn't specifically examine crude oil; other studies[28] converge in indicating that by 2050, oil supply will have fallen to a level sufficient to support no more than about half the world's current population in its present style of life. The challenges of adapting to the new circumstances are considerable, but it is not difficult to identify some key measures that could be adopted—such as adopting oil depletion protocols, reducing the amount of energy we waste, turning to renewable energy sources, strengthening communities through local food and local currencies, and reducing population.

Is this a doomsday message? Not necessarily. A more benign age may dawn for the survivors, in which they have more respect for themselves, for each other, and above all for the limits imposed by nature.[29]

dead plants before the slow sedimentation process has a chance to bury them underground. So the coal we extract and burn is the result of a unique period of Earth's history.

Oil and gas mainly formed from deposits sedimenting at the bottom of bodies of water by the decay of algae or plankton under low-oxygen conditions at some depth and in the absence of strong currents. The first result of this process is the mineral called "kerogen" (its name derives from the Greek word for "wax"), by far the largest stock of organic carbon in the Earth's crust. Humans have little interest in kerogen as a mineral, but it is of enormous importance because it is from kerogen that crude oil and natural gas are formed. Various stages of degradation may lead to crude oil and the degradation sometimes continues all the way to the simplest possible hydrocarbon compound, methane, which is the main component of natural gas. Normally, oil and gas can be found together in the same wells, with gas "capping" the oil reservoir below. In many cases, however, natural gas can also be found alone.

The special conditions that endowed the Earth with fossil hydrocarbons only existed in special periods of the remote past. We are extracting crude oil and natural gas mainly from deposits that formed in two specific periods, 90 and 150 million years ago, in the Mesozoic era.[30] We don't know exactly why these periods were so favorable for oil and gas formation, but it may be that their considerably hotter climate led to low-oxygen conditions in the coastal regions of shallow seas. Indeed, if you look at the distribution of oil fields today, you see that they are often aligned, probably as the result of the shape of ancient coastlines.

Not all oil existing today was formed during these two periods; some oil deposits are much older than the Mesozoic ones, while some even predate the Phanerozoic eon. And some relatively modern oil deposits date back just to the Cenozoic. But Mesozoic oil is prevalent, and since the Mesozoic era is known as the age of dinosaurs, oil is sometimes described as "dinosaur juice." This is a colorful image, but no more than that. The liquid that fills the tanks of our cars may perhaps contain traces of an occasional marine dinosaur, but the organisms that formed oil were by far microorganisms.

One specific characteristic of conventional oil and natural gas is that they are both fluids, gaseous or liquid, that are less dense than average crustal material, and so they tend to move upward as the result of hydrostatic pressure (also known as "Archimedes' principle"). If the rock that contains oil and gas is sufficiently porous, they can migrate to the surface, where they form pools that are slowly degraded by bacteria and oxidized to carbon dioxide. So, in order to accumulate underground in forms that can be extracted by human beings, oil and gas must be "trapped" in some way and moved to a porous reservoir capped by a nonporous "seal" that prevents them from migrating to the surface.

That scenario—a reservoir of oil or gas trapped in an underground reservoir and capped by a nonporous seal—in large part describes regular conventional hydrocarbon resources. But a vast range of nonconventional resources are also increasingly being exploited. These resources often exist in hard-to-reach locations, like the ultra-deep oil that was being extracted by the Deepwater Horizon, the offshore oil rig operated by BP that ended up spewing millions of barrels of oil into the surrounding sea (see figure 1.6). But the most commonly mentioned nonconventional oil resources are related to "shales," sedimentary rock that contains variable amounts of kerogen and, sometimes, crude oil and gas. The kerogen trapped in sedimentary rock is a combustible substance that can be transformed into liquid oil, but the process is complex and expensive. A similar problem exists with extracting oil from tar sands: oil that has been exposed to the atmosphere for long periods and has degraded into highly viscous bitumen. Tar sands are a solid that can be processed and transformed into liquid oil, but again, it is an expensive process. On this frontier, there has recently been much interest in shale gas and shale oil, liquids or gases trapped in shales that must be fractured ("fracked") to allow them to move easily to the surface. This is another expensive and complex process that turns out also to be destructive for the environment (see "Fracking: The Boom and Its Consequences," page 55).

Nearly all geologists today accept the fact that oil and gas have organic origins. However, the so-called abiotic theory—the fringe opinion that oil and gas are mainly the result of ancient inorganic processes—was considered as a

FIGURE 1.6. The Deepwater Horizon, the offshore oil rig at the epicenter of the BP disaster, was harvesting ultra-deep oil, one of the nonconventional resources increasingly exploited by the fossil-fuel industry.

possible explanation for the origin of oil in the early times of petroleum geology, and today it occasionally reaches scientific journals and becomes part of scientific debates. More frequently it appears on the Internet as an argument against the idea that the reserves of fossil hydrocarbons are limited. Proponents often claim that large (sometimes "immense") amounts of oil exist underground, mainly inside the mantle. Some contend that these primordial stores date back to the formation of the Earth. Supporters of this theory frequently claim that by tapping this gigantic reservoir we would never run out of oil, and the fact that we haven't done that, so far, is only the result of a conspiracy on the part of oil companies to keep the price of gasoline high.

But the abiotic theory completely fails to account for a number of critical characteristics of oil and gas, in particular their isotopic composition, which clearly indicates an organic origin.[31] Besides, the "immense reservoir" posed by the theory simply cannot exist, since it would contradict all that we know of the history of our planet and of its structure. Any oil present underground is continuously pushed toward the surface. If this primordial oil had formed billions of years ago, as the proponents of the abiotic theory maintain, it would have had all the time needed to find its way to the surface with volcanic eruptions at continental edges and at mid-ocean ridges. Once at the surface, it would have been oxidized long ago by bacterial processes and transformed into CO_2, consuming all the available atmospheric oxygen in the process. The Proterozoic oxygen revolution would never have taken place and Earth would have a reducing atmosphere mainly formed of hydrocarbons and CO_2, like that of Titan, the moon of Saturn. As it is normally expressed in the media and on the Web, the "abiotic theory" of oil formation is nothing more than an urban legend.[32]

The Death of Gaia

Mineral ores in the Earth are the result of an evolution that began billions of years ago, when the Earth was geologically more active than it is today. That evolution remains an ongoing process, although it progresses at slower rates than in earlier ages. Every mine we exploit today is Gaia's gift. But it is a gift that was made only once to humans, and when we have totally squandered it, it will be lost forever. The phase of mining by humans is a spectacular but very brief episode in the geological history of the planet. None of the minerals that we have so liberally dispersed all over the world will re-form in time for humans to use them again, even if we expected our civilization to last for tens or hundreds of thousands of years longer. No matter how long humans inhabit Earth, the genesis of new deposits and new ores is going to continue at the slow pace it has maintained for the past billion years. Slowly, most of the minerals mined and dispersed by humans will sediment on the bottom of oceans. These sediments will form a metal-enriched layer not unlike the iridium-enriched layer we see today in many places of the Earth—one that was formed, most likely, by the debris from the giant asteroid that hit the Earth 65 billion years ago, when the dinosaurs vanished from the scene. If, tens of millions of years from now, there are geologists living on the Earth, they may well find a metal layer corresponding to the mass extinction that is taking place today and wonder what caused it. They won't find any evidence that it can be related to an asteroidal impact.

Not all the metals that we have extracted and dispersed will remain buried in a sedimented layer. In tens or hundreds of millions of years, a large fraction of what we've dug from the Earth will have been transported by the oceanic conveyor belt to the edges of continents and recycled into the mantle. Part of it will have returned to the surface in the form of ores and deposits. Perhaps, if intelligent creatures exist in such a remote future, they will be able once more to mine the Earth's crust and create a new industrial civilization.

But not all mineral deposits will be re-created. Some of the ores that humans extract today, diamonds and coal among them, are the result of conditions that existed only in remote times.

There will be more irreversible transformations of the Earth's systems in the remote future. The convective movements of the mantle are expected to maintain the dance of the continents for a long time. If so, a few hundred million years from now, the present continents will coalesce and form a new supercontinent. The oceans will continue to be slowly absorbed into the mantle and be gradually reduced in volume, but they won't disappear before much more drastic events will have affected the Earth's ecosystem. On the same time scales, the sun will continue its gradual increase in luminosity, and that will increasingly strain the ability of the feedback mechanisms at the core of the Gaia system to keep the Earth's temperature cool enough to maintain liquid water on the surface. So far, the system has been fighting the sun's increasing heat by the gradual reduction in the concentration of CO_2, but there is a limit to how low that concentration can go. If it falls below a certain point, photosynthesis cannot be sustained, and without photosynthesis life as we know it cannot exist. Eventually, there is no escape from the fact that the increasing solar irradiation will cause the collapse of the Earth's ecosystem. All life will be extinguished; in time, the oceans will boil away and the Earth will become a hot and dry planet that will be eventually engulfed and destroyed by the expanding sun during its last phase of life, about five billion years from now.

The events of such a remote future do not concern us much, but we can imagine this outcome because we understand the mechanisms that have created the Earth's ecosystem as we know it today. That has led us to understand that the ecosystem is fragile. We know that the planet may continue to exist for several hundred million years, but we can't be sure of it, and we know that it must eventually die. In a sense, Gaia is getting old, and her life span might be much shorter than what we can theoretically calculate.

So, when it comes to mineral depletion, it is not just a question of asking for how long we can keep plundering the planet, but whether the planet—and its ecosystem—can survive the wounds we are inflicting upon it.

The excavators at the Garzweiler coal mine in Germany are among the largest machines ever built to move on land, lending a perspective on the size of modern mining operations.

2

Plundering the Planet:

The History of Mining

We could say that mining is as old as civilization, but really it is as old as life itself. All living creatures are miners, since all creatures need minerals that derive from the ecosphere, the thin shell of air, water, and rock at the surface of our planet. The body structure and metabolism of every living being are reported to depend on at least 16 chemical elements, but that estimate is likely low. Some sources specify a total of 26 elements, and others even 60,[1] although the role of some ultra-trace minerals is not yet clear. In any case, living creatures continuously acquire and exchange elements from their surroundings.

These elements come mainly from the atmosphere and are returned to it. Such is the case with the four basic elements supporting life: carbon, oxygen, hydrogen, and nitrogen. But in addition, living beings need phosphorus and calcium for their bones, sulfur for some of their amino acids and proteins, iron for transporting oxygen in the blood, and sodium for transmitting electric signals in the nerves and brain, to name just a few. These elements do not normally exist in gaseous form, and so they must be acquired from the Earth's crust.

For billions of years, life on Earth existed as single-celled organisms living in bodies of water. These creatures could obtain the elements they needed from ions dissolved in the water. Land plants appeared only around 350 million years ago. Their roots were a major evolutionary innovation that allowed the plants to mine the ground, absorbing from it the minerals they needed. As miners, land plants never went to great depths, limiting their action to the layer of fertile soil, at most a few meters thick, from which they could absorb mineral ions dissolved in water. But land plants have been very efficient miners. During the past few hundred million years the Earth has seen ice ages, hot ages, giant asteroid impacts, volcanic mega-eruptions, and other dramatic events, but on average the activity of the biosphere on the continents never changed too much. Land plants kept extracting minerals from the ground, and animals kept taking minerals from the plants. Everything was then reabsorbed and recycled in that immense chemical laboratory that is the humus layer, the true "skin" of planet Earth.

Today, it is estimated that the land biosphere produces 56 billion tons of new biomass every year.[2] Of the elements that are part of this mass, most come from the atmosphere, but about 1 percent must be extracted from the ground. Therefore, plants are mining about half a billion tons of materials from the crust every year. The cycle is very efficient: plants have never been in danger of "running out" of minerals at the planetary scale.

But recently something has changed. It has happened at an extremely fast rate, relative to the geological time scale. A species belonging to the animal kingdom has started to do something that no animal ever did before: to extract minerals directly from the ground, without the need for plants as intermediate providers. It is a species that digs, drills, crushes, extracts, and processes the ground into the mineral substances it needs. It is a species of miners: human beings.

About two and a half million years ago our ancestors started picking up stones from the ground for use as cutting and crushing tools. It was a slow start for an activity that, today, has become a major planetary force. Nowadays humans extract from the ground several billion tons of materials every year. We use all 88 of the elements present in the Earth's crust and even unstable elements that didn't exist on Earth in measurable amounts before we started creating them. We dig at depths unthinkable for plant roots: our mines are hundreds of meters deep and our drills reach tens of kilometers into the crust, even under the sea.

The activity of human miners isn't limited to making holes in the ground; it is changing the structure and the composition of the Earth's surface and atmosphere. Mountains are demolished and new ones are created. Elements and compounds that had been buried in the depth of the crust for hundreds of millions, even billions, of years are extracted and dispersed all over the surface. The composition of the atmosphere is changing with the increasing concentration of greenhouse gases, mainly carbon dioxide (CO_2), generated from the combustion of fossil fuels. We have even changed the composition and the structure of the very soil that we depend on for crops (see "Soil Fertility and Human Survival").

Humans are transforming the Earth into a different planet. How did we become miners on such a gigantic scale? It is a story that needs to be told from the beginning.

Origins of an Industry

Our remote ancestors started their career as miners simply by collecting rocks they found on the ground and using them to make tools. That simple act ushered in the Stone Age, which covers perhaps 99 percent of human history,

Soil Fertility and Human Survival

Toufic El Asmar

Perhaps our most important source of minerals can be found in the rich, complex ecosystem that blankets most of the Earth's land surface: soil. This all-important organic matter was formed over thousands of years as rock broke down into tiny particles that were gradually infiltrated by living organisms. Running anywhere between a few centimeters and several meters deep, soil sustains a diverse mix of plants and animals that forever change it as they live and die. It is moved about by wind, water, ice, and gravity—sometimes slowly, sometimes rapidly. And as history has shown us, it can make or break civilizations.

It's little surprise that many ancient civilizations began where the topsoil was richest and farming was most productive. But many of these civilizations mismanaged the soil, and as their agricultural productivity declined, so did their civilizations. Occasionally they vanished entirely. Studies suggest that the 1,700-year-old Mayan civilization in South America collapsed around 900 CE because its fertile ground eroded away due to bad soil management.[3]

Soil and survival are so intricately entwined because fertile soil supplies most of the elements that higher plants need to support photosynthesis and other metabolic processes. Only carbon, oxygen, and hydrogen come from water and air. Plants must absorb all the other elements directly or indirectly from the soil (or through artificial fertilization when their concentration in the soil is insufficient).[4] To grow, plants depend on especially large amounts of mineral nutrients such as nitrogen, phosphorus, and potassium and smaller amounts of the secondary minerals calcium, magnesium, and sulfur. They also depend on micronutrients such as boron, copper, chlorine, iron, manganese, molybdenum, and zinc. These micronutrients occur in very small amounts in both soils and plants, but their role is critical; a deficiency in one or more of them can lead to severe reduction in growth, yield, and crop quality.

Soil fertility is a complex process that involves the constant cycling of nutrients between organic and inorganic forms—something achieved through water, nitrogen, and carbon cycles and mediated by the nematodes, earthworms, bacteria, fungi, and other flora and fauna present in the soil. As plant and animal wastes decompose, they release nutrients to the soil. These nutrients may then undergo further transformations, mostly aided by soil microorganisms. Natural processes also bring changes: lightning strikes may fix atmospheric nitrogen in the soil by converting it to nitrogen dioxide; flooding can cut off the soil's supply of oxygen from the air, allowing denitrifying bacteria to convert the soil's nitrate into gaseous nitrogen, which can then escape the soil.

Yet even though the soil changes, the layers of an undisturbed soil will stay much the same during one human lifetime. However, when the soil is moved, scraped, or plowed, it can be destroyed in almost no time at all, particularly if land quality and land use are mismatched.[5] When the components that contribute to fertility are removed and not replaced, and the conditions that support soil fertility are not maintained, soil depletion occurs. This loss of fertility leads to poor agricultural yields, or even zero yields, especially in the case of crops, which are extremely sensitive to nutrient depletion.

Soil depletion occurs in many ways. In agriculture, depletion can result from excessively intense cultivation and inadequate soil management. For instance, in tropical zones where the nutrient content of soils is low, widespread soil depletion has resulted from overtilling (which damages the soil structure), insufficient nutrient inputs (which leads to mining of the soil's nutrient bank), and salinization. The combined effects of growing population density, large-scale industrial logging, slash-and-burn agriculture, ranching, and other factors have in some places reduced soil fertility to nearly zero.

In fact, billions of tons of soil are being physically lost each year. The most serious losses arise from erosion—the washing or blowing away of surface soil, sometimes down to bedrock. While some erosion takes place naturally, without human help, natural soil loss and new soil creation normally stay in balance. However, the rates of soil erosion associated with agricultural practices are accelerating, to the point of exceeding soil-loss tolerances over most of the Earth's cropland regions.[6]

The irrigation systems that have played an important role in increasing crop production have also had negative impacts on soil quality, with some researchers estimating that excessive watering has caused salinization. As figure 2.1 shows, the Food and Agriculture Organization (FAO) of the United Nations estimates that 34 million hectares (Mha), or 11 percent of irrigated areas, are affected by some level of salinization, with China, the United States, and India representing more than 60 percent (21 Mha) of the total impacted land. An additional 60 to 80 Mha are affected to some extent by waterlogging and related salinity.[7] The uncontrolled application of chemical and industrial wastes has degraded soil as well.

Not all soil loss is from farming, though. Millions of hectares of what would otherwise be good farmland are being flooded for reservoirs or paved over for highways, airports, parking lots, and expanding urban areas. Agriculture is also experiencing rising competition from fast-growing cities and urban settlements, resulting in smaller areas of productive agricultural land at a time when world population is growing and expectations are rising among people everywhere for a better life. Global warming, too, is expected to increase the rate of nutrient loss in soils, since microbial decomposition occurs faster under warmer temperatures.[8]

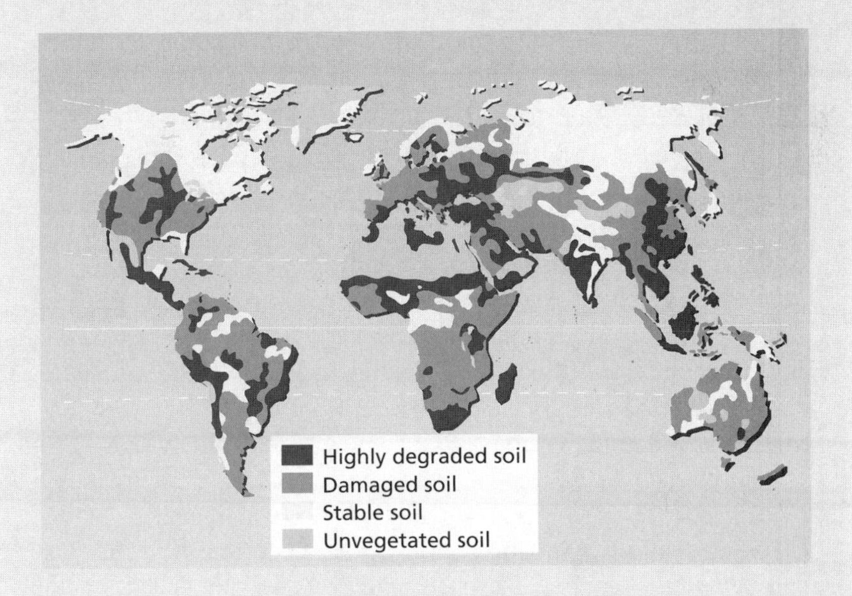

FIGURE 2.1. The state of soil health globally.

The Impact on Food Supply

The world is facing a series of challenges to human survival. Water is growing increasingly scarce, water pollution is becoming more widespread, and water-related ecosystems are degrading. Global warming, air and land pollution, and the depletion of natural and mineral resources are escalating. These are all serious threats to human welfare, but the loss of suitable land and soil quality for agricultural production is no less important and no less serious.

The summer of 2012 was the second hottest and driest since 2000.[9] Drought reduced grain production in the United States, wheat production in Russia, and agricultural production in most southern European countries, and especially Italy and Spain, badly hitting farmers already in trouble because of the increasing costs of fertilizers and fuels. As the Earth Policy Institute put it, "With prices rising, many of the world's poorer families had already reduced their consumption to one meal a day. But unfortunately for many families, even this is no longer possible. Millions of households now routinely schedule foodless days each week—days when they will not eat at all."[10]

The total land area of the world exceeds 13.2 billion hectares, but less than half of it can be used for agriculture, including grazing. The remainder is either too wet or too dry, too shallow or too rocky. (The single most serious drawback to farming additional land is generally lack of water.) In addition, some land is toxic, some is deficient in the nutrients that plants require, and some is permanently frozen. Europe, Central America, and North America have the highest proportion of soils suitable for farming, although a number of the more developed countries seem intent on paving over much of their best farmland with roads and buildings. The lowest proportions of arable soils are in north and central Asia, South America, and Australia. Around the world, but especially in developing countries, increasing competition for land and water has spurred a land grab, with state and commercial investors rushing to acquire tracts of farmland.

A report of the Natural Resources Conservation Service of the US Department of Agriculture showed that:

- Some of the world's land productivity has declined by 50 percent.

- Desertification can be observed on 33 percent of the global land surface and affects more than one billion people, half of whom live in Africa.
- Crop yield reduction in Africa due to past soil erosion may range between 2 and 40 percent, with a mean total loss of 8.2 percent for the continent.[11]

The report estimated that in 2001 southern Asia lost an estimated 36 million tons, or $5.4 billion, of cereal production to water erosion and $1.8 billion to wind erosion. On a global scale the annual loss of 75 billion tons of soil costs the world about $400 billion per year.

New Solutions

Unfortunately, there are no simple solutions to these gigantic, complex problems. We cannot expect that technology will come to the rescue with some miracle crop. The so-called green revolution that took place during the second half of the 20th century did increase crop yields, but in the process it used large amounts of artificial fertilizers and crops that required increased amounts of pesticides in order to survive. The productivity of the land is limited by basic factors such as the efficiency of natural photosynthesis, which cannot be modified by humans—not even by using fancy GMO crops.

We must recognize that we are in a state of deep overshoot for practically all the natural resources available to us. Agriculture is not an exception, even though it is theoretically renewable. What we are facing may be no different from the fate of many civilizations of the past. When farm productivity declined, society attempted to maintain production by expanding the land base under cultivation and putting more effort into cultivating the depleted areas. That led to accelerated soil loss, which became a major factor in the collapse of entire civilizations—such as the Mayan one. Without a significant change of paradigm in agriculture, our destiny will be the same.

So our approach to halting the decline of agriculture must be different. New agricultural practices must produce more food on less land by using fewer inputs and conserving and enhancing natural resources and biodiversity. Sometimes called save-and-grow farming, ecological agriculture, or sustainable crop production intensification, these practices draw on nature's contribution to crop growth—soil organic matter, water flow regulation, pollination, and natural predation of pests—and apply appropriate external inputs at the right time, in the right amounts. They also offer proven productivity and economic and environmental benefits. A review of agricultural development in 57 low-income countries found that eco-friendly farming led to average yield increases of almost 80 percent. It can also help mitigate climate change by sequestering millions of tons of carbon a year in soil.[12] But we also don't want to return to the ancient agricultural practices that required the work of large numbers of people who lived in conditions of poverty and exploitation that today we judge unacceptable. To avoid that fate, modern renewable energy technology may replace the energy supply that today comes from fossil fuels, but without the environmental costs and the depletion problems of fossil fuels.[13] Eventually, we'll learn how to cultivate the land without destroying it.

starting some 2.5 million years ago. Not just any stone could be a cutting implement—only those hard stones that could be chipped (or "knapped") to form a cutting edge would do. That limited the choice to flint, chert, and obsidian, a volcanic glass. Other kinds of stone, for instance jadeite, started being used as tools only in relatively recent times. These stones provided extremely sharp edges and could be transformed into deadly weapons for hunting and warring. (It is also possible that human males used them to shave their beards, though the jagged edge of knapped stone wouldn't have made the task easy.) The early times of human exploitation of minerals also saw other uses for rocks. Some, like pyrite and other forms of iron sulfide, were found to be able to create sparks when struck against hard rocks. That made lighting a fire much easier, especially compared to the laborious procedure of rubbing wooden sticks against each other until their friction generated enough heat to spark a flame. Other minerals, like ocher, were crushed and roasted to make pigments. Ocher was obtained from the iron mineral that we now call hematite; its name derives from a Greek word for blood to indicate its reddish (or, at times, yellow) color. It was used as a cosmetic, for body painting, possibly as medicine, and, surely, for painting on cave walls. Rocks were also used as throwing weapons, as counterweights for javelins, for boiling water after having been heated on a fire, as ornaments, as supports for lamps burning animal fat, and probably for much more.

It is possible that at some point our remote ancestors found themselves running out of the "easy" stones—those that could be found simply on the ground. It couldn't have been too difficult for them to understand that there were more stones of the right kind underground. But how to reach them? Stone Age humans didn't have tools that allowed them to dig much deeper than plant roots. But in time they found that some kinds of limestone can be broken even with simple tools, such as a deer horn. So, about 40,000 years ago, people started digging to find hematite and flint.[14] It was the true start of mining. Today, in England, we can still see ancient mines dug into white limestone. Most were dug around 10,000 years ago. These mines look much like their modern counterparts: deep underground tunnels where ancient miners laboriously crawled in search of minerals. Exploring these mines, modern-day archaeologists found that the tunnel roofs were still darkened by smoke from the miners' oil lamps. They found deer antlers the miners had used as digging tools. In some tunnels they even found human remains, perhaps miners killed by a collapse or human sacrifices to the dark deities of the deep.

These ancient mines appear not just in England but in several areas of northeastern Europe—in Belgium, Holland, Denmark, and wherever limestone chalk existed. Their earliest appearance has been dated to the transition period between the Paleolithic age and the Neolithic one, which saw the development of agriculture, pottery, statuary, and more. The Neolithic was a period

FIGURE 2.2. Prehistoric flintstone knife.

of intense technological evolution and rapid population increase. Mining, too, saw a rapid evolution that brought with it the age of metals. The mining of metals initiated a major technological step forward because metals, unlike stones, require complex processing methods to transform them into useful materials. It is likely that metallurgy was the first form of inorganic chemical processing practiced by humans.

The ancient history of metal discovery is complex and based on sparse evidence, and there are various opinions on when various metals were discovered and used.[15] But what is certain is that the ancients knew and used at least seven metals: gold, copper, silver, lead, tin, iron, and mercury. They may actually have used a few more. In particular, iron in those times was often alloyed with nickel, but ancient blacksmiths never identified nickel as a separate material. Zinc was used in ancient times, especially in alloys with copper, but very often it was confused with tin. Arsenic is also found as a component of copper alloys, but it is unlikely that the ancients knew it in its elemental form. Some ancient artifacts of antimony have been found in Egypt, but they are extremely rare. Finally, it has been claimed that the Chinese plated their bronze weapons with chromium in the third century BCE, the time of the terra-cotta army.[16] That, however, is unlikely, as it would have required sophisticated electrochemical technologies, surely not available at that time.

Of the seven common metals of antiquity, gold was likely the first to be extracted and used, marking the very beginning of human metallurgy. Though rare in the Earth's crust, gold used to be relatively common in alluvial deposits in rivers thanks to an erosion process caused by the long-term flow of water over gold veins. This "placer" gold was relatively easy to find; all that was needed was a certain eye for the areas of a river where gold nuggets would most likely accumulate. Once found, nuggets could be recovered by hand or by panning to separate the higher-density gold from ordinary riverbed stones. Panning required only a flat container, which would be agitated in such a way to keep the gold inside, while expelling the other wet sediments.

It seems that the first large-scale gold panning took place in a wide region of Europe, including the Carpathian region and the Balkans, during the fifth millennium BCE. In any case, the appearance of gold associated with human settlements starts squarely within the Neolithic period. There is no evidence that Paleolithic humans ever collected it, though doing so wouldn't have required technologies they didn't possess. Evidently the interest in gold was more a cultural factor than a technological one.

Once recovered in nugget form, gold can be formed into wires and sheet simply by hammering. However, some form of high-temperature processing

FIGURE 2.3. The Mask of Agamemnon, made from thin gold sheet, may date back to the 15th century BCE. Some doubts have been cast on the authenticity of this artifact,[17] but it and other similar masks give us some idea of the mastery attained by the earliest goldsmiths, as well as some of the most ancient realistic portraits ever discovered.

is necessary to fuse small nuggets together. Since gold melts at a temperature slightly higher than 1,000°C, it is impossible to do gold metallurgy in an open fire. So gold mining progressed in parallel with relatively sophisticated technologies to make high-temperature furnaces in which vigorous bellowing in of air increased the temperature of the burning charcoal.

At first pure gold was rare, since gold nuggets were usually found in the form of electrum, a natural pale yellow alloy of gold and silver. It seems that early metallurgists either could not or did not want to separate the gold from the silver. Pure gold and silver came much later in the Neolithic, when technologies were developed to remove silver from mined gold. But in the first millennium BCE, electrum was still in use for jewelry and coins. A major discovery of these ancient times was that pure silver could be obtained by processing lead minerals. With these developments gold and silver became relatively common all over the world.

Copper in metallic form came into use at about the same time as gold, although according to some reports it may have been used even earlier, as far back as 9000 BCE, in the region now known as Iran.[18] Copper can occasionally be found in its native form as nuggets of pure metal in alluvial deposits, and it is likely this form of copper that was first used by humans. In the early times of copper metallurgy, nuggets were hammered into shape or heated and welded together by hammering in a technology similar to iron forging, a process that would be developed much later.[19] With time, people developed high-temperature furnaces able to fuse and cast copper into homogeneous objects. Unlike gold and silver, which had only decorative uses in ancient times, copper was hard enough that it could be used as a tool, such as a hammer or a blade. These copper tools were much softer than the older stone tools and couldn't provide sharp and long-lasting cutting edges. But copper tools had the great advantage that they wouldn't break to pieces when used with excessive force. A copper axe with a yew handle was found with the frozen remains of Ötzi, the mummy of the Similaun glacier, at the border between Italy and Austria. This axe had been cast by the man who carried it, a fact revealed by the residues of copper and arsenic found in his hair. Ötzi lived at about 3300 BCE.

Soon the demand for native copper, much rarer than native gold and with multiple uses, grew beyond the supply, and new sources had to be found. In the Mediterranean region, the island of Cyprus became the center of copper production. It is likely that the ancient Cypriots first exploited the resources of native copper and, later on, found that metallic copper could be obtained simply by heating copper carbonates (malachite and azurite) and copper sulfides (chalcopyrite and others) in open furnaces. Heat can decompose carbonates, while the reaction with the oxygen of the air can remove sulfur, transforming these compounds into pure metals.

But pure copper was to be just a step in a progression toward more and more sophisticated metal tools. Soon it was discovered that copper and tin could be combined to form bronze. This alloy was much harder than pure copper and, as is the case with most alloys, it had a melting point lower than that of either component, so it was easier to melt and cast in useful shapes. But where could early metallurgists find the tin they needed to make bronze? Tin was much rarer than any of the other elements so far exploited, and it required a remarkable effort to extract enough to feed the new industry. When tin

FIGURE 2.4. A vase from the fifth century BCE shows a woman looking at her reflection in a mirror. That mirror was most likely made from bronze.

mineral resources were found in Cornwall, England, the area quickly became a major supplier for all of Europe. Tin was also found in Brittany and northwestern Spain, which led to the development of a trade system that brought tin all the way from northeastern Europe to the Mediterranean. This made bronze relatively common, though still expensive.

Bronze was used for a variety of purposes. Metal razors replaced the older generation of obsidian razors, which had never really been very practical. For the first time in history men could shave with ease! Then bronze was used for mirrors. The concept was not new; humans had been gazing upon their own reflections in the waters of rivers and streams for a long time, as the ancient Greek myth of Narcissus tells us. During the Neolithic, people made portable mirrors from obsidian and polished stone, but it was only with the arrival of copper and bronze that humans could have mirrors not unlike the ones we have today. Polished copper surfaces provided a reasonably good reflection, but it was soon discovered that a high concentration of tin alloyed with copper could make a clear reflecting surface. This alloy, often called mirror metal, stayed in use for millennia, up until the 19th century in Europe. Only in relatively recent times have mirrors been made by coating glass with a silver (and later aluminum) layer.

During classic antiquity, bronze had become cheap and abundant enough that it could be used to cast human-sized statuary and even larger pieces. Many of these ancient works of art have survived to the present day nearly intact. But bronze came to be used for much more aggressive purposes after it was discovered that it was an excellent material for weapons. In the early days of the age of metals, daggers had been made of pure copper, and some were long enough that they could be termed "swords." But copper was not hard enough that it could be used to make a real sword—for that, the strength and the toughness of bronze was needed.

The first bronze swords arrived with the second millennium BCE. Their blades were leaf-shaped and up to 90 centimeters long. With sharp points, they seem to have been mainly thrusting weapons designed to puncture the enemy's body. Bronze was also fashioned into shields and armor, and the Egyptians left us impressive images of the Sea Peoples armed with sharp-pointed bronze swords and equipped with shields and plumed helmets (see figure 2.5). Compared to this new generation of weapons, Ötzi's copper axe was just a child's toy. Thus dawned a new age of war that pitted professional fighters, clad in heavy armor and using deadly weaponry, against all those who couldn't afford this kind of equipment. It was perhaps the start of the distinction between nobles and commoners, which didn't exist in ancient tribal societies.

Then there was lead. It was obtained mainly from galena, a compound of lead and sulfur that was rather common around the Mediterranean Sea and is reported to have been mined and smelted for the first time in Anatolia, during the early years of human civilization. Being soft, lead didn't make a good blade, but human invention found ways to use it for war nevertheless. Lead weights made javelins more deadly, and ancient slingers shot lead projectiles—a way of killing people at a distance thousands of years before lead was fashioned into balls, then bullets, for firearms. Outside of warfare, lead had a very useful characteristic: it didn't harden when worked at room temperature. So it was perfect for pipes and vessels. Lead is so soft that for some purposes it may have been simply chewed into shape, as was still being done into relatively modern times.[20] Its low melting point (at a little more than 300°C) made it easy to cast into objects like figurines, the precursors to today's toy soldiers, now

FIGURE 2.5. This ancient Egyptian illustration shows a battle between Egyptians and Sea Peoples, who hold swords most likely made of bronze. These thrusting weapons were probably the most lethal weapons of the time.

made of less toxic materials. Lead's great versatility spawned uses in everyday items like dishes and cups. It is said that the Roman Empire fell because the Romans poisoned themselves by drinking wine from leaden cups. In reality the fall of the Roman Empire is a much more complex story,[21] but it is true that the Romans used lead for food processing and storage and may never have realized the danger they put themselves in. We may not have been much smarter than the Romans, having used lead as a gasoline additive and blown it up into the atmosphere for decades.

Mercury was introduced to the ancient inventory of metals relatively late, when it was discovered that it could be obtained by roasting the mineral cinnabar, a form of mercury sulfide. Long before it was used to make mercury metal, cinnabar was used as pigment because of its spectacular red color, which gives it the common name vermilion. And five thousand years ago vermilion was being used to preserve human bones in Neolithic burials in Spain.[22] Mercury metal was a different matter: it must have been a big surprise to discover that heating the bright red crystal of cinnabar in an open furnace generated a shiny metal that remained liquid at room temperature—a characteristic that made mercury unique among the metals known to the ancient.

Despite the fascination it inspired, mercury didn't have many practical uses in ancient times. It was used as cosmetic and as a medicine but was especially unsuitable for both purposes, being highly poisonous. Mainly it was used as a reactant for noble metal metallurgy, since mercury can dissolve most metals, forming an amalgam—a liquid or semiliquid alloy. So, if mercury were put in contact with gold-containing minerals, even those with such tiny amounts of gold that it wasn't visible to the naked eye, it could dissolve the gold, thereby extracting it from the mineral mass. Then, by heating the amalgam, it was possible to vaporize the mercury and recover the gold. The same procedure could be used to recover silver. It was also possible to use a gold-mercury amalgam to plate metal objects with gold, a technology that was used for thousands of years. That method of gold plating has been abandoned today because mercury vapor is extremely toxic, but we can imagine the fate of many ancient goldsmiths who used it. It was perhaps because of its intimate relationship with gold that ancient alchemists thought that mercury could be key to finding the "philosopher's stone," the legendary substance that could create gold from other metals (and even, said some, restore health or bestow immortality). It didn't work, though, and all they achieved was poisoning themselves by mercury fumes.

Of all the metals used in antiquity, one is seen as most important: iron. "Cold iron, the master of them all," as Rudyard Kipling said in one of his poems. Iron didn't have the luster of silver and gold, nor their "nobility"—that is, their resistance to oxidation. Iron was dark in color and rusted easily, but it was to become the most commonly used metal in the world, a characteristic that it maintains to this day.

The first manufactured iron objects appear in the archaeological record about midway through the second millennium BCE. In these early times, metallic iron was found mainly in meteorites, either in pure form or alloyed with nickel. As such, it didn't need much processing, and it could be hammered into shape when either cold or, better, red hot. But meteorites, although not rare, were difficult to find. The only practical way to detect one was to look on ice sheets or in desert areas, where the color contrast between the meteorite and the ice (or the sand) made it easy to spot. It seems that meteoric iron was very rare and more expensive than bronze, so much so that ancient records describe it as if it were a precious metal. In addition, before the complex technologies needed to transform iron into steel were developed, bronze was a superior material in terms of strength and hardness. So for centuries iron remained rare, scarcely used for practical purposes.

The eventual switch from bronze to iron may have been forced by the disruption of the tin trade in the Mediterranean region caused by the migrations of the Sea Peoples, the multiethnic tribes who raided Egypt and other areas in ancient times.[23] The disruption to trade may have been a problem at first, but eventually it ushered in a major technological revolution in the extraction and use of iron. One of the reasons iron plays a major role in human history is that it is found in great abundance in the Earth's crust. But transforming those abundant iron minerals into metallic iron was much more difficult than any operation that early metallurgists had attempted before.

The problem with iron metallurgy is that, as a metal, iron is very reactive toward atmospheric oxygen. Copper sulfides, for example, can be transformed into metal just by heating them in the presence of air, but doing the same with iron would result only in transforming iron sulfides into oxides—useless for metallurgy. Iron oxides transform into metallic iron only if heated in the presence of charcoal in a low-oxygen atmosphere. In these conditions, the carbon of the charcoal reacts with the oxygen of the oxide and disappears in the form of a gaseous compound (CO_2), leaving metallic iron in the furnace. But that environment was not easy to create with the technologies available to the ancient metallurgists. Working in an open furnace, as had been the rule up to then, was out of the question, because oxygen in the air would simply burn all the carbon in charcoal before it could react with the iron oxide. But a closed furnace still needs oxygen to burn charcoal to reach high temperatures. Finding the right conditions and building the right kind of furnace—one that would reach high temperatures without needing so much oxygen—was an art more than a science in an age where there was no way to precisely measure temperatures or gas composition. But the techniques for performing this delicate operation were gradually perfected, and the abundance of iron ores made metal tools become relatively cheap and available to everyone.

However, even with the best furnaces available in ancient times, iron could not be melted and cast in the same way that copper and bronze could. Its melting point was too high, despite the great efforts made by ancient blacksmiths to increase temperatures in their charcoal-fired furnaces. What these furnaces could normally do was to create a "cake" of semi-molten iron, which also contained residual oxides and impurities ("slag"). This cake had to undergo a second stage of processing, when it was heated again to near melting temperatures and then laboriously hammered into shape—a process that also removed the slag. This necessity of forging iron launched the trade of the blacksmith, a trade that has figured prominently in human history for thousands of years.

The problem with iron produced in this way is that the resulting material is soft, softer than cold-worked bronze. To transform it into something that can be sharpened and used as a blade, iron needs to be turned into steel, a material that we know today to be an alloy of carbon and iron. While the atomic properties of steel weren't understood until the 20th century, people in ancient times had discovered by trial and error that if they could manage to add some carbon to iron and then quench the resulting alloy, they obtained a hard material that could be used to make excellent swords—much better than anything that could be done with bronze (and better razors, too!).

However, making steel added another layer of difficulty to iron metallurgy. Adding carbon to iron with the equipment available to ancient blacksmiths was very difficult since the oxygen of air tended to burn away all the carbon that the blacksmith laboriously tried to add to iron. The best that most ancient blacksmiths could do was to squeeze a little carbon into the outermost part of the blade. This made it dead hard outside and soft inside—a fine cutting tool as long as it wasn't strained too much. But in battle such a sword would easily lose its cutting edge and bend, to be restraightened with a knee and a prayer. That's the reason ancient swords were usually so thick and heavy.

The saga of steel spans millennia and includes many legends. One held that in order to make good steel it was necessary to quench the red-hot sword in the body of a live slave. We don't know whether that practice was intended in allegoric terms or whether people were actually sacrificed in the belief that their death would give to the sword some kind of supernatural properties. But the existence of this and other legends illustrates the great difficulty that early blacksmiths had in making steel. As late as the Middle Ages, good steel-making technologies were still not available in Europe, although they existed in the Middle East and in Asia. The famed swords of Damascus, for example, were made with high-carbon steel developed and produced in India.[24]

With all their problems, however, decent steel swords came rather cheap in comparison with the old bronze ones and could be used to equip large armies. The society that perhaps made best use of steel started as a humble village in central Italy: Rome. The Romans' warlike society grew by gobbling up its neighbors one by one. Soon they became experts in iron metallurgy. The fact that Romans would not normally sport a beard was a fashion, in part, but also a message about their technological ability to make steel. For the Romans, being well shaved meant saying to their enemies, "Be careful, we have sharp blades!" As miners the Romans surpassed their old teachers, the Etruscans, developing highly sophisticated technologies for their times. In order to remove large amounts of rock from an excavation site, they would heat it with fire and then flood it with cold water to crack it. It was then easier to remove the weakened rock using picks. With these methods the Romans extracted not only iron but also gold and silver from their mines in Spain. On the other side of Eurasia, the Chinese had also developed good mining technologies and created an empire based on iron weapons.

These seven common metals of antiquity continued to be extracted and used until the fall of the Western Roman Empire in the fifth century CE. With that, Europe entered the Middle Ages badly depleted in minerals.

Fossil Fuels and the Birth of Modern Mining

The Middle Ages began as a period of great hardship in Europe but eventually led to a new age of mining. Black powder, which had been invented in China probably as early as in the ninth century CE, was imported to (or perhaps rediscovered in) Europe a few centuries later. It changed not only the way wars were waged but also the way mines were operated. Mining became a literally explosive activity. With black powder, crushing rock and digging tunnels became much easier, and mines began to look the way we think of them today: deep excavations and long underground tunnels dug into rock.

With this new technology it was possible to reactivate old mines and restart mining in Europe. But the real revolution in mining was the discovery of the New World by Europeans. That put an end not only to the Middle Ages but also to mineral scarcity in Europe, as the "virgin" American continents had never been exploited for their mineral resources. The new abundance came first with precious metals, gold and silver, which led to mineral rushes during which the native inhabitants were ruthlessly exterminated or enslaved. By the 16th century Spanish adventurers such as Hernán Cortés and Francisco Pizarro had devastated and destroyed the Aztec and Inca empires, all in the name of gold. The gold and silver of the American continents became the source of power of the Spanish empire. For centuries the Spanish exploited the famed Cerro Rico ("rich mountain") of Potosí in Bolivia as a source of silver, condemning the local laborers to horrendous work conditions. Gold rushes took place in many more places, the most famous being the California gold rush of 1849, with its related saga of the forty-niners, devastation of the land, and extermination of the native inhabitants.[25] It was neither the first nor the last case of rapid production growth in a new mining region, with a subsequent quick decline. The newly discovered continents of the era—North America, South America, and Australia—were to become new sources of all the traditional metals that had been exploited in Europe in ancient times.

The evolution of mining at the end of the Middle Ages was not just a question of new regions to be exploited. The 18th century saw a true avalanche of discoveries in chemistry. Several involved the identification and the separation of new metals. During this century the seven common metals of antiquity were joined by 16 new metals, including cobalt, chromium, platinum, zirconium, and uranium.[26] Several nonmetals were also isolated in the 17th and 18th centuries, including hydrogen, oxygen, chlorine, and phosphorus.

The 19th century saw the continuation of this rapid rhythm of discovery, which led Dmitri Mendeleev to create the periodic table in 1869 to systematize the world's knowledge about the elements. In the 20th century researchers moved on to the identification of unstable radioactive elements, and in the 21st century we are still trying to extend the range of known elements to extremely unstable isotopes with half-lives of fractions of milliseconds.

But the real change was not with the mining of these new elements. Rather, it came with the exploitation of a very well-known one: carbon, in the form of fossil fuels. This new source of energy gave an incredible boost to society and created the world we know today.

It all started with coal. It is reported that the Romans were the first to use coal as fuel, exploiting the abundant resources they had access to in Britain, while the Chinese were already burning coal by the 13th century, as we can read in Marco Polo's *Il Milione*. But true large-scale coal mining started only during the 18th century in Europe, and in particular in England and France. Initially coal was considered a poor fuel, but with the development of coking (baking coal to burn off impurities), mineral coal could be used for the same tasks as wood charcoal, but at a much lower price. That changed many things. For instance, for most of human history iron had been smelted with charcoal, which made it such an expensive commodity that it was used to make little more than weapons and armor. Now, produced in coal-fired forges, it became so cheap that it was possible to make everyday items in iron, such as pots, pans, and more. In the 19th century iron columns, complete with ornate iron capitals that imitated the marble capitals of old Greek temples, became popular. Cheap coal also made steel cheap, allowing it to be used for a new generation of weapons, from cannons and muskets to "ironclad" battleships, which started being manufactured in the early 19th century.

Coal did more than make iron cheap; it powered the steam engine. The first steam engines were used to pump water out of coal mines. They were very inefficient, but it didn't matter. Coal was inexpensive and abundant. The pumps made it possible to extract more coal, and more coal could power more pumps, leading to more coal being extracted. With time, the steam engine became efficient enough that it could power ships and locomotives as well as factories. As William Stanley Jevons wrote in 1865, "Coal in truth stands not beside but entirely above all other commodities. It is the material energy of the country—the universal aid—the factor in everything we do. With coal almost any feat is possible or easy; without it we are thrown back into the laborious poverty of early times."[27]

With coal, Britain experienced the first industrial revolution. An awesome complex of factories, people, and machines became the inner powerhouse of the British empire. The idea spread quickly to other countries. France had started her coal revolution perhaps even earlier than Britain; in fact the French Revolution that started in 1789 was born from the need to get rid of the old landed aristocracy to make room for a new, coal-based economy. Germany, too, developed its national mines, and slowly the revolution spread to eastern Europe, to Poland and Russia, and later on to North America. But the domain of King Coal was not destined to last forever.

Coal was perhaps the first important mineral resource of modern times to show depletion problems. England's production peaked in the 1920s and was soon followed by Germany's. France would peak a couple of decades later, but without ever approaching the production magnitude that England and Germany had achieved. Coal had created the European world empires; its decline was to spell their demise. King Coal was abdicating, at least in Europe.

The history of coal didn't end with the decline of the European producers. The lead was picked up by new producers in North America, China, and Australia, and coal is now the fastest growing energy resource in the world. But the importance of coal was destined to decline anyway thanks to the appearance of a new mineral commodity: crude oil, which was more versatile, more powerful, and easier to transport.

The modern history of crude oil starts around the mid-19th century, and it had a very humble beginning. At that time coal powered almost everything, but not quite. There was a market niche that coal could not occupy: lighting. You could burn coal, but you couldn't make a practical coal-powered lamp. Indoor lighting was still the domain of an ancient technology that had accompanied humankind for millennia—oil lamps powered by vegetable oil or animal fat. The mid-19th century saw the development of a treatment that turned "rock oil" into kerosene, which worked well in lamps and generated great demand. That, in turn, led people to search for new sources of this kind of oil, which now was starting to be termed "crude oil." Up to then it had been obtained in limited quantities from surface pools or from oil that had seeped into the sea and could be collected with sponges. But it was the task of American prospector Edwin Drake to find the right way to increase the supply. In 1858, in Titusville, Pennsylvania, he was the first to drive a pipe into the ground to look for oil. He found it at a depth of 20 meters, and it was the start of a saga that is still ongoing today.

As it often happens, when something is available at low cost, new uses are found for it. Among the by-products of manufacturing kerosene was gasoline. It was a liquid too flammable and volatile to be used in lamps, and so it was sold as a cheap stain remover for clothes. At around the end of the 19th century it was found that gasoline could be used as a fuel for internal combustion engines. These engines had been around for quite a while, but they couldn't be made into practical devices for lack of suitable fuels. Gasoline changed all that, and the first four-stroke engine, invented by Nikolaus August Otto in 1876, was the start of a revolution. Later on, in 1892, Rudolf Diesel invented the engine that takes his name, which could use a different fraction of distilled crude oil, today called diesel fuel. With these engines, a vehicle could be light enough that it could travel on roads and compete with horse-driven carriages. It was a revolution in transportation.

Other uses for crude oil were also found, including in the expanding market for rubber for tires. At the time rubber was manufactured only from natural sources: tropical trees. But the increasing number of vehicles on the road generated a furious rush for rubber, to the extent that millions of people were killed in the Congo in the struggle for control over that region's rubber production.[28] Crude oil came again to the rescue, and starting with the first decades of the 20th century processes to make synthetic rubber from oil gradually became common and phased out natural rubber. At about the same time, it was found that a viscous form of crude oil, bitumen, could be used to pave roads, obtaining a smooth surface, perfect for the new rubber tires. The triad of the internal combustion engine, synthetic rubber tires, and paved roads led to road transportation as we know it today: cars and trucks everywhere.

Crude oil also made possible other transportation technologies. In 1903 the Wright brothers flew the first engine-powered airplane. It was the first step in the long development of aviation, which even today can exist only thanks to the high energy density of fuels obtained from crude oil. This trait also made crude oil interesting for naval applications. At the beginning of the 20th century, commercial and military ships started to switch from bulky and inefficient steam engines to steam turbines powered by either coal or oil. Later most vessels switched to diesel engines, which were more practical and reliable.

In the 1960s crude oil surpassed coal as the main source of energy for the world's economy. It was the start of a period of great prosperity, a level of wealth perhaps never before seen in history. For the Western world, it was the time of a car in every garage, a refrigerator in every kitchen, and a TV set in every living room. Cars sported fins that made them look like small spaceships. They were used for traveling, but also much more. Americans vacationed in cars, ate in cars, watched movies in cars, and . . . well, many of the present generation may have been conceived in the backseat of a car: sons and daughters of crude oil, so to speak. Crude oil also contributed to the conquest of space begun in the 1960s. It provided the energy for missiles and for the spaceship that would, in 1969, bring men to the moon for the first time in history.

The fossil-fuel revolution took place without almost anyone taking notice of the long-term cycle of oil extraction that was unfolding. Just a few people had occasionally voiced the idea that we were using finite resources and that, sooner or later, we would run into problems of depletion. But the early claims were not based on good data and were swept away by the onrushing wave of new discoveries that kept the mineral industry growing, fueled by the seemingly unlimited energy produced by fossil fuels.

Things changed in the 1950s, when good data about the world's hydrocarbon resources started to become available. At that time American geologist Marion King Hubbert started wondering how long it was possible to maintain the increasing rate of oil production that was the rule at that time. He developed a statistical

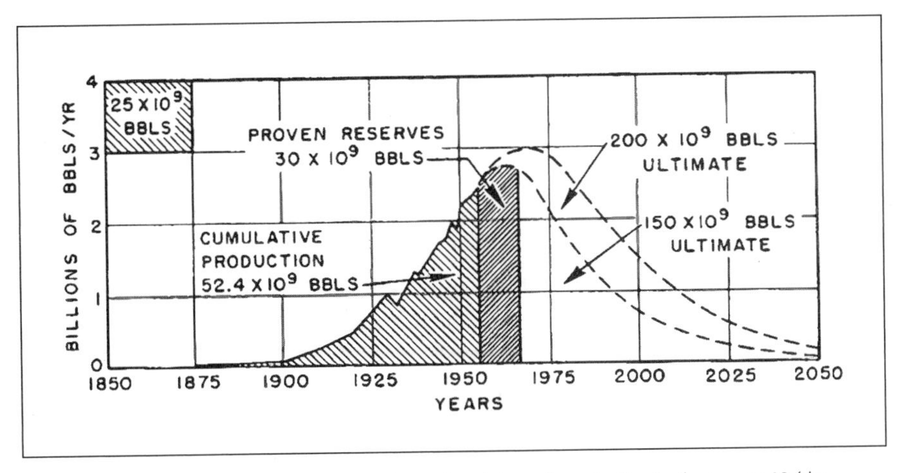

FIGURE 2.6. The Hubbert model accurately predicted that oil production in the Lower 48 (the continental United States excluding Alaska) would peak in the 1970s. The model described very well production in the United States until a few years ago, when new extractive technologies generated a new growth trend in production.

model that became popular and eventually took his name: the Hubbert model.[29] According to Hubbert's model, the maximum possible level of oil production in the United States should have arrived around 1970 (see figure 2.6). Many took his prediction as pure madness, but it came to pass with remarkable accuracy. The peaking and successive decline of production in the United States, up to then the world's largest producer, was not without consequences on the world stage. It was one of the elements that generated the first great oil crisis, which started in 1973.

The crisis took everyone by surprise, but it was not really unexpected. In the 1960s Pierre Wack, a Shell Oil analyst, had used a method called "scenario planning" to analyze the situation.[30] Up to then world oil production had been increasing at a nearly constant rate of 7 percent per year, but it was clear that, in order to keep going at that rate, enormous investments in new exploration and new infrastructure were needed. Wack noted that such investments weren't being made. Something had to give, and the result was the first oil crisis. The crisis lasted at least 10 years before production again started to increase in the mid-1980s, with the arrival on the market of oil from the North Sea and from renewed production facilities in Saudi Arabia. The world's oil production system never recovered the low prices and fast growth rate of precrisis times. Nevertheless, it gained a period of respite that lasted about 20 years. Then oil prices started going up again, reaching another peak in 2008 at a level of almost $150 per barrel to stabilize later on at a plateau of about $100 per barrel.

These events should not have been unexpected. Hubbert, in 1956, had already applied to the whole world his model for crude oil production, finding that troubles could be expected sometime around the turn of the century. More recent studies had estimated that the global peak of oil production—dubbed "peak oil" by Colin Campbell—would occur at some moment during the first or the second decade of the 21st century (see figure 2.8).[31] We still don't know for sure if this global peak has occurred, because it is masked by production oscillations generated by market factors. But we may be very close to it, as one of the expected consequences of the peak is a rapid price increase, which is what we have been seeing. If this is the case, in the coming years we are to see an epochal change as the world's oil production starts an irreversible decline.

With the slowdown of the growth trends for crude oil, natural gas has acquired more and more importance. In the past, the gas associated with oil wells was simply burned in place because it was too expensive to transport. However, with more difficult times arriving, it became convenient to develop ways to transport gas over long distances. The problem is that storing natural gas requires heavy, expensive pressurized vessels, and transporting it requires complex and expensive infrastructure. On land gas is transported through a

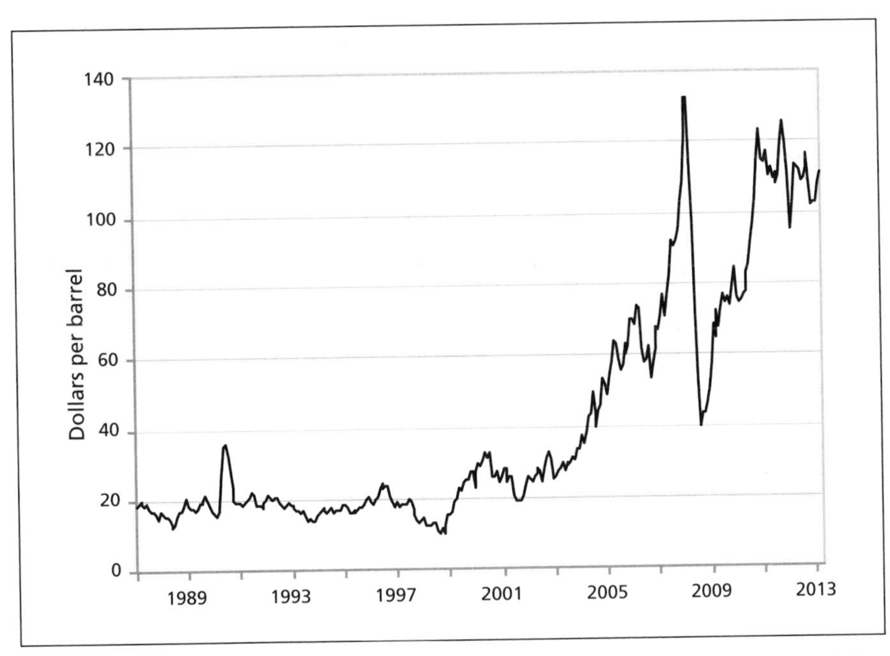

FIGURE 2.7. Oil prices from 2000 to 2012 show an increasing trend that became dramatic in 2008. Afterward prices declined for a brief period before soaring again. It is clear that we are seeing the effects of real problems with the extraction of oil, rather than simply the effects of speculation.

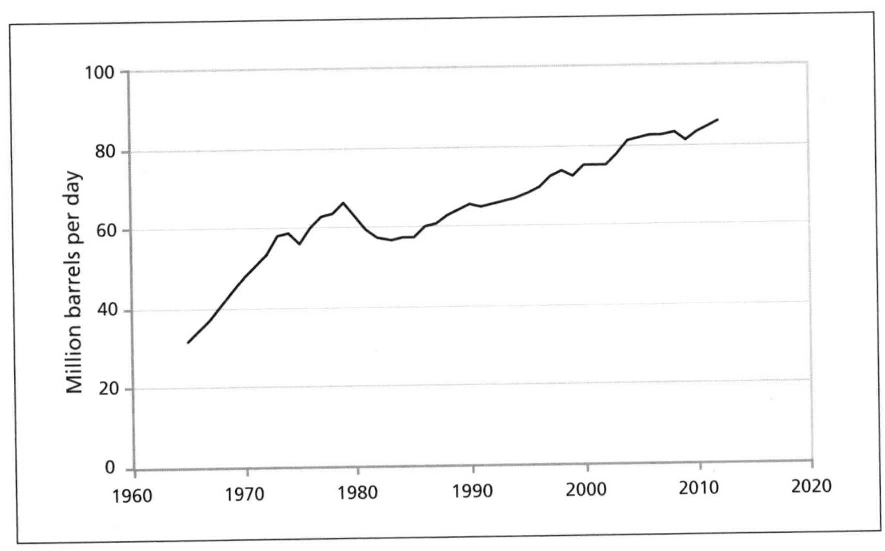

FIGURE 2.8. Recent world crude oil production trends. After a peak that took place in the 1970s, growth has been very slow and only recently surpassed the earlier peak. The most recent data indicate that crude oil production has reached a plateau that may be the symptom of an impending decline.

network of pipelines. To travel by sea, gas must undergo cryogenic liquefaction to obtain a sufficiently high-energy-density liquid (liquefied natural gas, or LNG) for transportation in special refrigerated tankers. These methods are far from being satisfactory: pipelines cannot cross oceans, and cryogenic transportation is expensive. So gas remains mainly a regional resource, and it makes little sense to speak of a global production peak for gas in the same way we would for oil. But local gas peaks are possible. Several have been observed, such as the 1971 peak in the United States.

However, in recent times the development of technologies to extract gas locked in shale deposits ("shale gas") has prompted a return to high levels of gas production, especially in the United States.[32] Because of this achievement, some have been speaking of a new era of prosperity based on shale gas.[33] Most likely that is an exaggeration. (See "From Shale Gas to Tar Sands Oil," page 197.) Shale gas, like all fossil fuels, is a limited and nonrenewable resource. So at some point in the future we are going to see a global peak for gas, although it may be fragmented into several regional peaks that occur at different moments.

The Short-Lived Cycle of Nuclear Energy

While fossil fuels were going through their cycle of growth and decline, another mineral resource took the attention of the world as a potential revolutionary factor in energy production: uranium. Whereas fossil fuels store solar energy accumulated in ancient biological processes, uranium stores energy created in the explosion of ancient supernovas. When a way to unleash this energy was found, it generated the first nuclear explosion at Alamogordo, New Mexico, in 1945 and then the destruction of Hiroshima and Nagasaki in Japan. Together with the first nuclear weapons, the first nuclear reactors were developed—initially for the purpose of creating the plutonium needed for the bombs, then with the additional ability to produce electric power. The enthusiasm for these discoveries was incredible. Many claimed humankind was entering an atomic age that would herald a period of nearly infinite prosperity.

The growth of the new technology was extremely rapid. By the 1980s the United States and the Soviet Union together had stockpiled something like 70,000 nuclear warheads,[34] a good illustration—if ever there was one—of the concept of "overkill." Civilian nuclear reactors also saw rapid growth from the 1950s up to the 1980s, when industry growth leveled off. From then on, fewer new plants were built. The existing ones are aging, and the nuclear industry is facing an unavoidable decline.

Fracking: The Boom and Its Consequences

Ian T. Dunlop

Fracking, or more accurately, hydraulic fracturing, is the process of opening up fractures in tight subterranean geological formations by injecting fluid at high pressure. Fracking was initially developed by the oil and gas industry in the late 1940s and has since been widely applied.[35] Today over 50 percent of conventional oil and gas wells around the world are fracked. Most of these reservoirs are relatively localized and lie far below the surface, where fracking does not interfere with other critical activities, such as agriculture or water aquifers. As a result, conventional oil-field fracking has not spurred widespread concern.

More recently, though, fracking has gained notoriety as it has been applied increasingly to the production of oil and gas from so-called nonconventional resources in shale beds and coal seams. These deposits are closer to the surface, where fracking has the potential to fundamentally interfere with agriculture and water availability, create minor earthquakes and other geological disturbances, generate substantial pollution, and cause health risks.

Despite these grave concerns, fracking for nonconventional resources has expanded enormously in recent time, driven by higher oil and gas prices. Traditional fracking technology has been adapted to include techniques like directional (nonvertical) drilling, allowing extensive areas of shale beds and coal seams to be accessed in ways never previously possible. In the United States this has led to a significant increase in nonconventional oil and gas production, to the point where proponents argue that the country could become energy independent in a matter of years.[36]

Proposals are afoot for a massive expansion of fracking activity around the world, following the US experience. (See "From Shale Gas to Tar Sands Oil," page 197.) However, the downsides of fracking are now starting to emerge, with questions also raised about the viability of nonconventional production.[37]

How It Works

Fracking injects water, chemicals, and "proppants" (sand or sandlike components) into the Earth to increase the flow of other fluids, such as oil, gas, or water, from subterranean reservoirs.

These reservoirs typically exist in sandstone, limestone, shale, or other material with pores that can contain the fluids. The greater the number and size of the pores, the more fluid they can store. The more interconnections between the pores ("permeability"), the more easily the fluid can move under pressure toward an eventual collection point.[38]

Conventional oil and gas reservoirs are primarily sandstone or limestone with relatively high porosity and permeability. Fracking in this context is used to improve what are already fairly high flow rates. The process requires relatively small amounts of energy, fluid, and other materials.

Nonconventional oil and gas resources, on the other hand, are contained within reservoirs such as shale beds or coal seams, which have much lower porosity and permeability. In both shale and coal seam fracking, the process injects fracturing fluid under high pressure to create the fractures and carry proppant material into the formation to keep the fractures open. In contrast to conventional reservoirs, the tightness of nonconventional reservoirs means that high volumes of fluid injection are required to be successful.

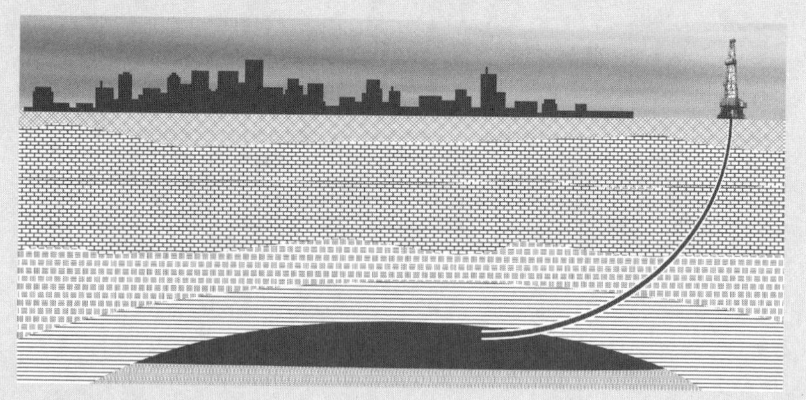

FIGURE 2.9. Schematic description of the fracking process, which involves crushing the rock around a horizontal drilling pipe in order to mobilize the gas locked inside the rock.

Wells are drilled either vertically or directionally and then lined with steel casing (see figure 2.9). After the fracturing injection is made, much of the fluid is circulated out of the well, necessitating surface storage facilities, which can often be large given the volumes of fluid involved, particularly in coal seams.

The Production Outlook

Unlike conventional oil and gas wells, shale beds and coal seams extend over wide areas. These reservoirs lack uniform oil and gas quality and content, so developers will try to access the "sweet spots" first. Typically the quality of the reservoir then deteriorates over time.

However, the greatest production challenge lies in the fact that shale and coal seam gas wells exhibit a "shooting star" production profile. Once fracking has been carried out, production rises rapidly to a peak, but it then declines rapidly, too, often by 80 to 95 percent over the first three years,[39] as the oil or gas around the fractured area is exhausted. As a result the countryside has to be peppered with wells to maintain the production required to provide a return on investment, often several thousand wells in a single shale play. Directional drilling can offset this problem by allowing some extended reservoir areas to be fracked from one well, but geographical spread remains an issue.[40]

Environment and Health Impacts

Conventional oil and gas production is localized and operates under high technical standards, which is essential given the acute safety issues involved. However, shale bed and coal seam development, given their geographical spread, are most unlikely to achieve those standards, a fact that can lead to degraded or diminished agricultural lands, contaminated water, depleted aquifers, and more gas leaks.[41] These potential risks have already caused significant conflict between the energy and agricultural industries in the United States, Canada, and Australia, and significant protest from environmental groups and citizens living near fracking operations.

The large amount of water consumed in nonconventional fracking is a particular concern. Nonconventional extraction has been booming in Pennsylvania's Marcellus Shale, but it requires around 15 million liters per frack, far more than the 0.4 million liters historically required to frack a conventional well in the same area.[42] This enormous amount of water often has to be moved by truck, or possibly by pipeline. The dramatic water consumption is a concern everywhere this type of fracking occurs, but it is particularly an issue in arid regions, like the western United States and Australia.

Post-fracking, the recovered water and other fluids are largely toxic waste. They have to be treated in surface facilities and eventually disposed of, something that often happens by reinjecting the fluids into dry gas wells. The recovered gas has to be moved by pipeline to link to major distribution pipeline systems.

All this, in multiple locations across vast frack sites, creates many additional opportunities for contaminating surrounding air, land, groundwater, and surface water. It also poses a number of significant health concerns.

Fracking fluid is designed specifically for its target, and formulations are often proprietary. But these mixtures can contain numerous chemicals, many of them carcinogenic, which can pose serious risks to humans, animals, and the environment in general. Efforts to force the industry to disclose the chemicals abound, and in some areas disclosure is now mandatory.

In high-density drilling areas, ozone levels appear to be rising dramatically, prompting concerns about respiratory health problems and lung disease. Many regions with increased fracking are also experiencing earthquakes, microseismic events brought on as intense hydraulic pressure and fluid movements impact subterranean formations. However, arguably the greatest environmental risk is fracking's impact on climate change.

The Climate Connection

Our collective refusal over the last 20 years to reduce emissions means the world can only burn less than 20 percent of existing proven fossil fuel reserves if catastrophic climate change is to be avoided.[43] This fact removes any justification for the continued development of fossil fuel resources, conventional or nonconventional, unless carbon capture and storage (CCS) is available to sequester their emissions—and there is no sign that this will happen either in the time or to the extent required to avoid catastrophic climate change.

As Nobuo Tanaka, former executive director of the International Energy Agency (IEA), said when launching the agency's "Golden Age of Gas" report in 2011: "While natural gas is the cleanest fossil fuel, it is still a fossil fuel. . . . An expansion of gas use alone is no panacea for climate change."[44]

So the fact that fracking allows us to tap additional fossil fuel reserves is a serious issue in itself. But the actual process of fracking presents other serious climate challenges as well. The most critical one results from what are called fugitive emissions, which are basically uncombusted gas leaks. These can occur when natural seepage to the surface is exacerbated by fracking pressures, or from leaking facilities spread across the countryside and not maintained to adequate standards, or from venting gas to the atmosphere.

The gas is primarily methane. When burned, it has roughly half the emissions of coal. But if methane is leaked to the atmosphere before combustion, it has a warming potential around 25 times that of CO_2 over a 100-year time frame, or 72 times over a 20-year time frame, the latter being more relevant given the current rapid acceleration of climate change. A leakage rate of around 3 percent negates the advantage gas has over coal from an equivalent warming perspective. Typical shale and coal seam gas leakage rates were thought to be in the 1.5 to 2.5 percent range, but recent evidence suggests that they are considerably higher.

The refusal by industry proponents and governments to take climate change seriously has meant that reliable baseline measurements of fugitive emissions (like baselines for water and land quality) have rarely been undertaken, as they were not thought to be important. This makes it extremely difficult to determine the real climate impact of fracking for nonconventional gas, but the emerging measurements are not encouraging.

Natural gas is often touted as a more climate-friendly alternative to coal, but even this appears suspect. Coal burning emits not just carbon but also aerosols, tiny particles that are suspended in the atmosphere and have a cooling effect. Cleaner-burning gas does not emit aerosols, so if coal use drops, to be replaced by gas, the level of aerosols drops correspondingly and hence the cooling effect is reduced. So one consequence of the wider use of gas is likely to be an increase in global temperature for several decades, due to the removal of the aerosols that are currently holding temperatures at a lower level than would otherwise be the case.

The combination of aerosol reduction and fugitive emissions means that fracked nonconventional gas almost certainly has a worse warming impact than coal, possibly far more so.

Moreover, as highlighted by the IEA, extensive fracking, by disturbing subterranean formations, may limit the potential for secure CCS, and CCS is an essential technology for the continued use of gas.[45]

Incentives and Disincentives for Fracking

The costs of fracking are considerably higher than those of conventional oil and gas production. The incentive for fracking is that the availability of cheaper hydrocarbon resources is declining, just as demand increases with increasing population. The resulting oil price increase, coupled with the fact that substantial nonconventional hydrocarbon resources were discovered in the United States, provided the trigger for the technological innovations that led to their development. Given the long-standing US goal to wean itself from dependence on foreign oil and gas, it was inevitable that fracking would take off, negative environmental impacts and climate change concerns notwithstanding. It is also no surprise that fracking is generating great interest globally, given US experience and the extensive shale and coal resources available in other parts of the world.

However, fracking has to be put into a broader context. It demonstrates the rapidly declining energy return on energy invested (EROEI), which is evident with all nonconventional hydrocarbon resource development.

As figure 2.10 shows, the average EROEI required to run industrial society as we know it is around 8 to 10. Shale gas and coal seam gas are both at, or below, that level if their full costs are accounted for, as are shale oil and tar sands.[46] Thus fracking, in energy terms, will not provide a source on which to develop sustainable global society.

Fracking has been a remarkable story of innovation, built off basic oil and gas industry technology. At the same time it has major environmental disadvantages quite apart from its likely climate change impact, disadvantages that will become even more acute as the global

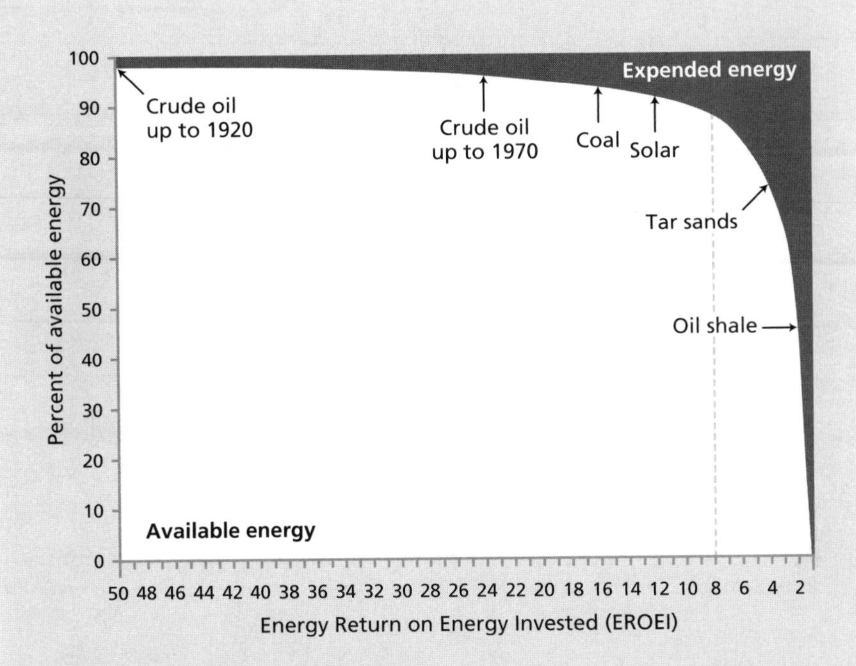

FIGURE 2.10. This image shows the high density of perforations needed to exploit a fracking area.

conflict between fuel and food intensifies. Overall, it is a classic case of increasing technological complexity leading to diminishing returns.

The IEA recognized this quandary in its May 2012 report *Golden Rules for a Golden Age of Gas*, emphasizing that without high environmental standards and the minimization of greenhouse gas emissions throughout the entire natural gas supply chain, the industry's "social license to operate" would be withdrawn.

At a time when we need to move away from fossil fuels altogether, a global rush into fracking is certainly unwise and will waste substantial funds that would be far better spent on low-carbon alternatives.

Several reasons led to the nuclear industry's decline, but the main problem may have been a strategic one: the need to stop the proliferation of nuclear weapons. Today, decades beyond the Cold War, nuclear proliferation is becoming a less pressing concern—except for those states defined as rogue, which

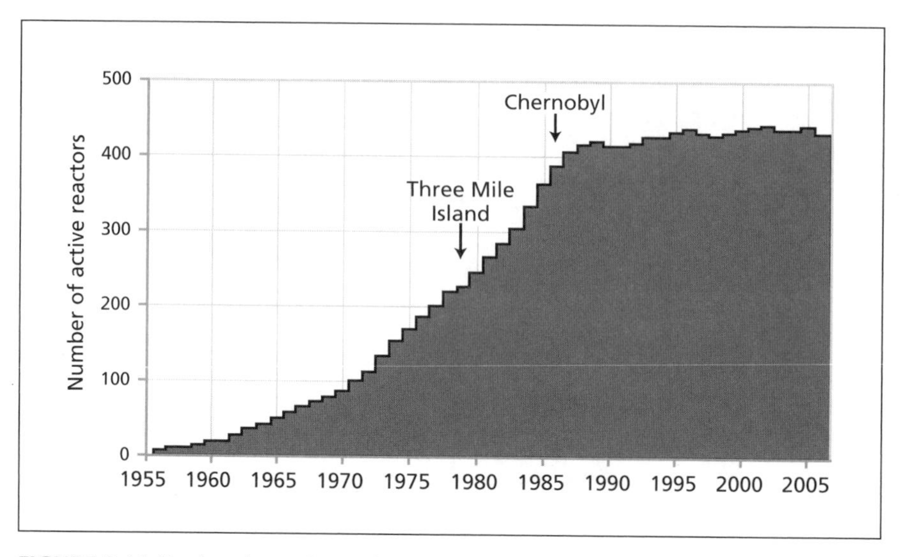

FIGURE 2.11. Number of operating nuclear reactors in the world. The initially rapid growth of nuclear energy tapered off in the 1970s, and it is now in decline.

seem to be most keen to acquire nuclear technology. So we are seeing a rebirth in the interest in nuclear energy. But there remains a basic problem: uranium is a mineral resource, and as such it exists in finite amounts.

Even as early as the 1950s it was clear that the known uranium resources were not sufficient to fuel the "atomic age" for a period longer than a few decades. That gave rise to the idea of "breeding" fissile plutonium fuel from the more abundant, non-fissile isotope 238 of uranium. It was a very ambitious idea: fuel the industrial system with an element that doesn't exist in measurable amounts on Earth but would be created by humans expressly for their own purposes. The concept gave rise to dreams of a plutonium-based economy. This ambitious plan was never really put into practice, though, at least not in the form that was envisioned in the 1950s and '60s. Several attempts were made to build breeder reactors in the 1970s, but the technology was found to be expensive, difficult to manage, and prone to failure. Besides, it posed unsolvable strategic problems in terms of the proliferation of fissile materials that could be used to build atomic weapons. The idea was thoroughly abandoned in the 1970s, when the US Senate enacted a law that forbade the reprocessing of spent nuclear fuel.[47] A similar fate was encountered by another idea that involved "breeding" a nuclear fuel from a naturally existing element—thorium. The concept involved transforming the 232 isotope of thorium into the fissile 233 isotope of uranium, which then could be used as fuel for a nuclear reactor (or for nuclear warheads).[48] The idea was discussed at length during the heydays of the nuclear industry, and it is still discussed today; but so far, nothing has come out of it and the nuclear industry is still based on mineral uranium as fuel.

Today, the production of uranium from mines is insufficient to fuel the existing nuclear reactors. The gap between supply and demand for mineral uranium has been as large as almost 50 percent in the period between 1995 and 2005, but it has been gradually reduced during the past few years. The most recent data available show that mineral uranium accounts now for about 80 percent of the demand.[49] The gap is filled by uranium recovered from the stockpiles of the military industry and from the dismantling of old nuclear warheads. This turning of swords into plows is surely a good idea, but old nuclear weapons and military stocks are a finite resource and cannot be seen as a definitive solution to the problem of insufficient supply.

With the present stasis in uranium demand, it is possible that the production gap will be closed in a decade or so by increased mineral production. However, prospects are uncertain, as explained in "The End of Cheap Uranium." In particular, if nuclear energy were to see a worldwide expansion, it is hard to see how mineral production could satisfy the increasing uranium demand, given the gigantic investments that would be needed, which are unlikely to be possible in the present economically challenging times. At the same time, the effects of the 2011 incident at the Fukushima nuclear power plant are likely to negatively affect the prospects of growth for nuclear energy production, and with the concomitant reduced demand for uranium, the surviving reactors may have sufficient fuel to remain in operation for several decades. In any case, high costs, high risks, and dilemmas over how to store nuclear waste over the long term, coupled with supply uncertainties, make it appear unlikely that uranium will be able to play the role of a major new energy resource that was once attributed to it.

There remains, theoretically, a further mineral source of energy: hydrogen, the element that generates the energy of stars by nuclear fusion. A typical refrain of the so-called atomic age was that if we could reproduce the same phenomenon in a controlled way, here on Earth, we would have a practically infinite energy resource. Indeed, hydrogen is enormously abundant as a component of the water molecule in oceans. The problem is that fusing hydrogen atoms together in stars requires a combination of high pressures and high temperatures that is impossible to reach on Earth. Besides, stars are so bright because they are so big, but the fusion that goes on inside them is a very inefficient process.

The End of Cheap Uranium

Michael Dittmar

Debates about nuclear energy usually focus on its advantages, disadvantages, and risks—or on the unsolved problem of storing radioactive waste. Most discussions, though, fail to address an obvious question: Can enough uranium can be mined to maintain or even increase the role of nuclear energy in the world?

As it turns out, the answer is no. Mineral uranium resources cannot provide the nearly unlimited abundance of energy that proponents of nuclear energy sometimes describe. It's true that there are large quantities of uranium in the Earth's crust, but there are limited numbers of deposits that are concentrated enough to be profitably mined. If we tried to extract those less concentrated deposits, the mining process would require far more energy than the mined uranium could ultimately produce.

Fifty years after commercial nuclear-fission power began, nuclear reactors still produce less than 14 percent of the world's electric energy (20 percent in the richer OECD countries).[50] So even a minor shift from fossil fuels to nuclear power would require a huge effort to replace aging reactors and construct hundreds of new nuclear power plants during the next 20 to 30 years. It would also require a significant increase in the worldwide uranium supply.[51] Facts like these present important barriers even for maintaining today's small contribution of nuclear energy to the world energy mix.

Uranium Resources and Reserves

Large-scale uranium extraction started after the Second World War, initially as the result of the demand created by the nuclear arms race. The superpowers engaged in this race were also among the main producers of mineral uranium. The United States mined about 370 thousand metric tons during the last 50 years (peaking around 1981 at 17 thousand tons/year). The Soviet Union and Canada each mined about 450 thousand tons. By 2010 global cumulative production totaled about 2.5 million tons. Of this amount, 2 million tons was used for electric energy production and is not available any more. Of the rest, a relatively small fraction had been used to manufacture nuclear weapons, but half a million tons were stockpiled by the military, mainly in the United States and the Soviet Union.[52]

Western Europe offers a good example of uranium production's pattern of growth and decline. Even though minor amounts of uranium are still produced in some European countries, for all practical purposes the uranium mining cycle ended there during the 1990s, after a total of about 460 thousand tons had been extracted.[53] Today, almost all of the 21 thousand tons/year needed to fuel European nuclear plants must be imported.

The analysis of the European mining cycle allows us to determine how much of the originally estimated uranium reserves could be extracted before further mining was considered impractical. The data shown in table 2.1 are perhaps surprising in their consistency: in all countries where mining has stopped, it did so when an amount well below the initial estimates (typically 50 to 70 percent) had been extracted. Similar conclusions can be drawn about the eventual halt of uranium mining in South Africa and the United States—even though, in these cases, the mining cycle is not completely closed yet.

Table 2.1. Uranium Mining Data

Country	Demand in 2010*	Peak production (and year)*	Initial resource estimate*	Extracted total*	Percentage extracted
Germany	3.45	7.1 (1967)	334.5	219.5	66%
Czech Republic	0.68	3.0 (1960)	233.4	109.4	47%
France	9.22	3.4 (1988)	110.8	76.0	69%
Bulgaria	0.28	0.7 (1985–88)	49.1	16.4	33%
Hungary	0.30	0.6 (1960–83)	32.8	21.1	64%
Romania	0.18	2.0 (1956–58)	37.1	18.4	49%
Spain	1.46	0.3 (1994–2000)	26.4	5.0	19%
Western Europe as a whole	21	12.3 (1976)	810	460	58%

* Amounts given in thousand metric ton units

Modeling Future Uranium Supplies

Using historical data for countries and single mines, it is possible to create a model to project how much uranium will be extracted from existing reserves in the years to come.[54] The model is purely empirical and is based on the assumption that mining companies, when planning the extraction profile of a deposit, project their operations to coincide with the average lifetime of the expensive equipment and infrastructure it takes to mine uranium—about a decade. Gradually the extraction becomes more expensive as some equipment has to be replaced and the least costly resources are mined. As a consequence, both extraction and profits decline. Eventually the company stops exploiting the deposit and the mine closes. The model depends on both geological and economic constraints, but the fact that it has turned out to be valid for so many past cases shows that it is a good approximation of reality. This said, the model assumes the following points:

- Mine operators plan to operate the mine at a nearly constant production level on the basis of detailed geological studies and to manage extraction so that the plateau can be sustained for approximately 10 years.
- The total amount of extractable uranium is approximately the achieved (or planned) annual plateau value multiplied by 10.

Applying this model to well-documented mines in Canada and Australia, we arrive at amazingly correct results. For instance, in one case, the model predicted a total production of 319 ± 24 kilotons, which was very close to the 310 kilotons actually produced. So we can be reasonably confident that it can be applied to today's larger currently operating and planned uranium mines. Considering that the achieved plateau production from past operations was usually smaller than the one planned, this model probably overestimates the future production.

Table 2.2 summarizes the model's predictions for future uranium production, comparing those findings against forecasts from other groups and against two different potential future nuclear scenarios.

As you can see, the forecasts obtained by this model indicate substantial supply constraints in the coming decades—a considerably different picture from that presented by the other models, which

Table 2.2. Uranium Supply and Demand through 2030

Scenario	Production 2010 (ktons/year)	Forecast 2015 (ktons/year)	Forecast 2020 (ktons/year)	Forecast 2025 (ktons/year)	Forecast 2030 (ktons/year)
Demand +1%/year	68	71.5	75	79	83
Demand −1%/year	68	65	61	58	55
This model	53.7	58 ±4	56 ±5	54 ±5	41 ±5
WNA	53.7	70	80	85	70
EWG	53	63–65	68–72	70–88	65–84
Red Book from IAEA	70–75	96–122	98–141	80–129	75–119

Note: Several possible scenarios are shown in this table: a slow growth of 1 percent yearly for the nuclear industry; a slow decline of 1 percent yearly; the forecast of the model presented here ("this model"); the 2009 forecast of the World Nuclear Association (WNA);[55] the 2006 forecast of the Energy Watch Group (EWG);[56] and the forecast given by the International Atomic Energy Agency in its 2009 edition of *Uranium: Resources, Production and Demand*, commonly known as the Red Book (RB).[57] The model presented here shows that even if nuclear energy production is not expanded, it is not possible to fuel reactors without tapping military reserves. In other words, extraction alone will not meet demand.

predict larger supplies. The WNA's 2009 forecast differs from our model mainly by assuming that existing and future mines will have a lifetime of at least 20 years. As a result, the WNA predicts a production peak of 85 kilotons/year around the year 2025, about 10 years later than in the present model, followed by a steep decline to about 70 kilotons/year in 2030. Despite being relatively optimistic, the forecast by the WNA shows that the uranium production in 2030 would not be higher than it is now. In any case, the long deposit lifetime in the WNA model is inconsistent with the data from past uranium mines.

The 2006 estimate from the EWG was based on the Red Book 2005 RAR (reasonably assured resources) and IR (inferred resources) numbers. The EWG calculated an upper production limit based on the assumption that extraction can be increased according to demand until half of the RAR or at most half of the sum of the RAR and IR resources are used. That led the group to estimate a production peak around the year 2025. But as we have seen in the United States and South Africa, RAR numbers are inconsistent with actual production. It is reasonable to assume that the EWG study would be much more consistent with our forecast if realistic RAR data were available.

The largest upper production limit, with large uncertainties, comes from the Red Book capacity scenario. The Red Book authors acknowledge, however, that the capacity numbers provided by the different countries are unreliable and much larger than actual mining results.

The predictive power of different forecast models can be judged from their ability to describe past mining data. Currently the only model that matches actual results when applied to past data is the simple 10-year lifetime model described here

Understanding the Limits

With the wealth of historical data available, we can now say that there exists a worldwide upper production limit for uranium extraction and that a production decline from existing mines will be unavoidable during the present decade. Assuming that all planned uranium mines can be actually opened, annual mining will increase from today's level of 54 thousand tons/year to a maximum of about 58 (±4) thousand tons/year in 2015. After 2015 uranium mining will decline by about 500 tons/year up to 2025, and much faster thereafter. The resulting annual production is predicted to be 56 thousand tons/year in 2020, 54 thousand tons in 2025, and 41 thousand tons in 2030. Such uranium supply constraints will make it impossible to sustain a significant increase in the production of electrical power from nuclear plants in the coming decades.

In practice, past attempts to obtain controlled nuclear fusion as a source of energy had hinged on the possibility of fusing a heavier isotope of hydrogen, deuterium. But not even the controlled deuterium-deuterium reaction is considered feasible, and the current effort focuses on the reaction of a still heavier hydrogen isotope, tritium, with deuterium. Tritium is not a mineral resource, as it is so unstable that it doesn't exist on Earth. But it can be created by bombarding a lithium isotope, Li-6, with neutrons that in turn can be created by the

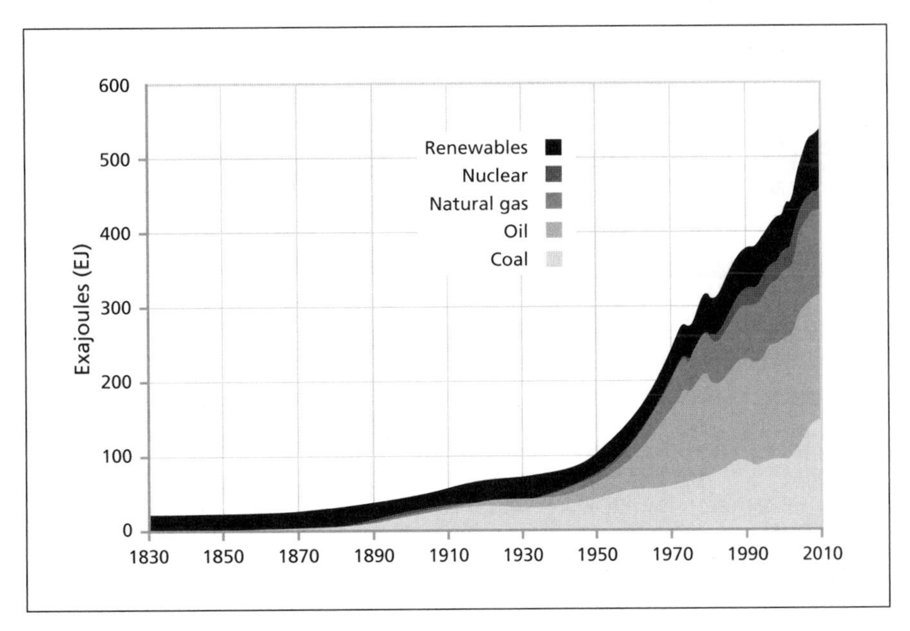

FIGURE 2.12. Comparison of the primary world sources of energy. Fossil fuels are still the main component of the mix. Nuclear energy is in decline, while renewables are rapidly growing but still far from catching up with the more traditional sources. In fact, the largest share of renewable energy remains today associated with an old technology: hydroelectric dams.

deuterium-tritium fusion reaction. (In this sense a fusion reactor is another kind of "breeder" reactor, as it produces its own fuel.) However, since the mineral resources of lithium are limited, and since the Li-6 isotope forms only 7.5 percent of the total, the problem of mineral depletion exists.[58] So not even nuclear fusion, if it were attainable, would give us the infinitely abundant energy described during the optimistic period of the atomic age.

That doesn't mean that nuclear fusion could not provide humankind with useful energy for a long time. Perhaps new mineral resources could be found to fuel power plants, such as the helium-3 isotope that is continually emitted in the solar wind and that might perhaps be collected in space. The problem with nuclear fusion is that several decades of efforts haven't yet led to anything that could even remotely produce energy, to say nothing about producing energy at costs compatible with what we can afford. The efforts in new prototypes are continuing, but there is no doubt that the atmosphere of general optimism about fusion in the atomic age is gone. That has led to a number of claims that fusion can be obtained in conditions much less extreme than those that the current understanding of nuclear physics tells us are necessary. It is the miracle of "cold fusion" that would put a small nuclear reactor on everyone's desktop.[59] Unfortunately, the possibility of using these devices for energy production has never been demonstrated. Many of the claims in this area are only the result of a "pathological science" approach,[60] lacking sufficient rigor and reproducibility. In short, if we want energy from nuclear fusion reaction, our best bet is to use the one fusion reactor that we know to be working and that we already have: the sun.

A Giant Industry in Continuing Evolution

Despite the projections of future decline, today the production cycle for mineral resources is far from being concluded. The world's industry is a voracious consumer of minerals, and it has been consuming more and more during the past two centuries. The amounts of minerals extracted nowadays is immense. Just for the United States, the available data indicate a grand total of about 3 billion tons per year. Figure 2.13 shows some data for the total minerals produced in 2010.

This amount becomes even larger if we consider the "extraction" of fertile soil in agriculture—consumed by erosion—as mining. It is estimated that about 4 billion tons of agricultural soil is eroded in the United States and dumped into the oceans every year.[61] Global estimates have ranged from 75 billion tons per year to 120 billion tons.[62] These amounts dwarf those created by natural erosion, which is at least one order of magnitude smaller.

To this amount we must add the amount of rock and sand moved by the construction industry. From US Geological Survey (USGS) data, we find that the worldwide production of sand and gravel may exceed 15 billion tons per year. The total world production of concrete in 2012 was more than 3.5 billion tons. China alone produces more than a billion tons—about 450 kilograms per person on average. According to geologist Bruce Wilkinson of Syracuse University in New York, we can visualize the total amount of rock and soil yearly moved by humans this way: "ca. 18,000 times that of the 1883 Krakatoa eruption in Indonesia, ca. 500 times the volume of the Bishop Tuff in California, and about 2 times the volume of Mount Fuji in Japan. At these rates, this amount of material would fill the Grand Canyon of Arizona in ca. 50 years."[63]

The production of sand and stone is usually referred to as quarrying. True mining, on the other hand, generally refers to other kinds of commodities, mainly metals. The USGS keeps data on the worldwide production of all minerals, and by far the largest metal commodity produced is iron, at almost 2 billion tons annually worldwide. So iron still holds first place among minerals,

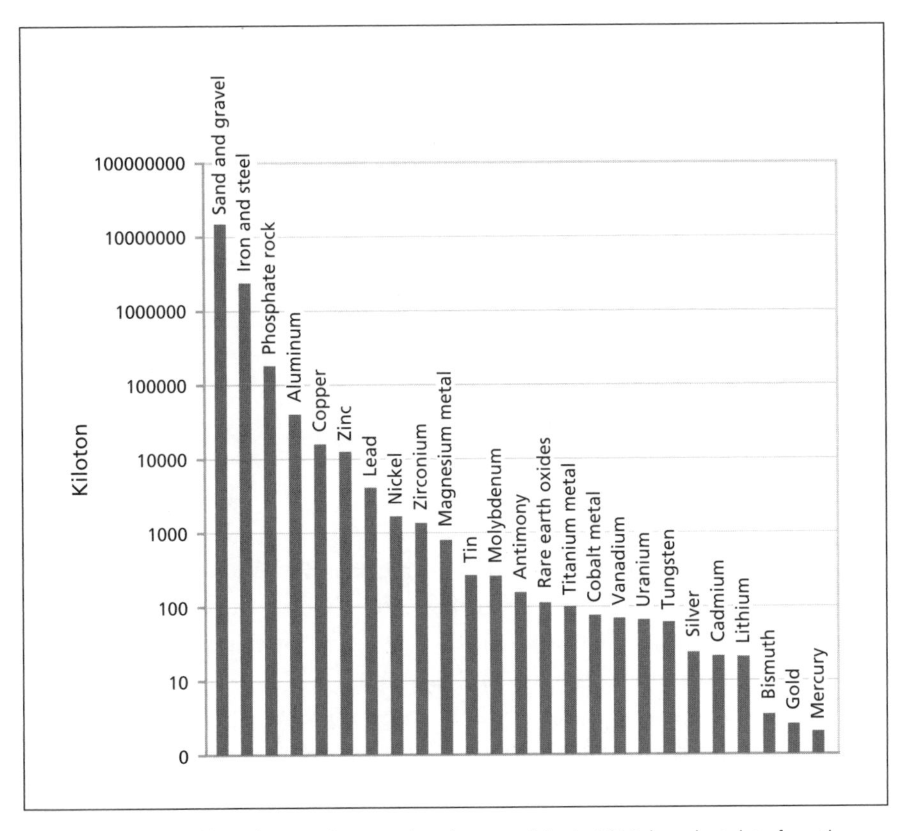

FIGURE 2.13. World production of some mineral commodities in 2010, based on data from the British Geological Survey.[64]

as it has since classical antiquity. Traditional metals, such as copper, are mined today in amounts much larger than anything possible in ancient times, with copper reaching 15 million tons per year. Other metals unknown in antiquity are also being mined and produced in large amounts.

By the 19th century, the ample supply of energy that came from coal made it possible to produce metals such as aluminum—abundant in the Earth's crust, but hard to obtain because it is highly reactive with oxygen and so could not be smelted in a charcoal furnace. Once electrochemical methods to reduce aluminum were developed, it became a major worldwide mineral commodity, with about 35 million tons per year produced today. Aluminum is rarely used alone; it is normally alloyed with other light elements to improve its strength and other characteristics. Copper-aluminum alloys are the main structural material used today in aeronautic and aerospace applications.

With time, further electropositive metals became available. Magnesium, lithium, beryllium, titanium, and others started to be produced in significant amounts during the 20th century, although none had the success of aluminum in terms of amounts produced and widespread applications. Magnesium was found to be slightly better than aluminum in terms of strength per unit weight, but also more sensitive to corrosion. So magnesium metal is used mostly as a minority component of aluminum alloys. It is used as a majority component only for niche applications where extreme low weight is needed, such as in sport cars and planes. Its world production is less than half a million tons per year.

Titanium turned out to be expensive to produce, but it had one characteristic in which it was superior to steel and all the other light metals: its ability to resist high temperatures without losing its mechanical properties, something that made it indispensable for a variety of applications, especially in aeronautics. Titanium also has the advantage of not corroding easily. But titanium remains expensive to produce, and today the world's production of titanium metal is only slightly more than 100,000 tons per year, according to USGS, much less than the 10 million tons per year of titanium oxide that is produced, for use mainly as pigment.

Among structural metals, beryllium showed great promise but had to be abandoned because it turned out to be highly toxic for human beings. Lithium is the lightest of the metals, and its alloys show promise as structural materials. However, like magnesium, it suffers from corrosion problems. Nevertheless, lithium has found an important market as a minority component in other light structural alloys. Lithium's most important application today is as the active element in a new generation of batteries that promise to revolutionize transportation. Even so, the USGS data indicate that lithium production remains very small in comparison to that of the other light metals, reaching just a little more than 30,000 tons in 2011.

Other elements of the periodic table are found in different applications and different markets. Among the semiconductors, silicon is surely the most important, with a world production of more than 5 million tons. Most silicon is used as a component of steel, but in its ultra-pure form silicon is the backbone of the electronics and photovoltaic industries.

Gold's importance as currency has faded over the past century, but its production remains large for a noble metal at about 2,000 tons per year, though it is dwarfed by the production of silver at more than 20,000 tons per year. Other noble metals have found applications in the chemical industry, mainly as catalysts. Platinum, palladium, and rhodium are the fundamental components of the common "three-way" catalytic converters for automotive engines. These metals are produced in amounts of just a few hundred tons per year.

Even smaller is the production of rare metals such as gallium, at less than 100 tons per year, which is indispensable for flat-screen TVs and has many applications in advanced electronics. Another rare metal for high-tech applications is indium, which is a fundamental component for transparent conductive layers on flat-screen displays, with just 500 tons produced annually worldwide. Tellurium, a component of a new generation of thin-film solar cells based on cadmium telluride, has a similar production level of about 500 tons per year. At the smallest end of the production scale we find the rare earth scandium, with only about 100 kilograms of the metallic form being produced per year[65] (a couple of tons of the oxide form are produced annually). Other rare earths are simply not produced in metallic form. In comparison, precious and semi-precious stones are produced in much larger amounts. For instance, diamonds are produced in amounts on the order of 20 to 30 tons per year, worldwide.[66]

As we see, the mineral industry is a vast and variegated world. Every mineral being produced has its history, its mines, its peculiarities, its market. This system generates the flux of mineral commodities that make the world's industrial system work and grow. The only way to feed the ravenous creature, so far, has been to continually invest more and more resources in the mining industry and upgrade it with more and more aggressive methods of extracting and processing.

Black powder, discovered during the Middle Ages, transformed the mining industry with its explosive power. A new generation of explosives was created in the 19th century: in 1840 Ascanio Sobrero invented nitroglycerin, an explosive much more powerful than black powder, and Alfred Nobel created dynamite in 1863, a true revolution in mining. Coupled with the ability to move large amounts of rock by diesel-powered machinery, dynamite generated mining as we know it today, mainly based on "open-pit" mining. No more tunnels, no more digging. Entire mountains are demolished by dynamite charges and then swept away to access the minerals inside. This method is especially destructive when used for coal mining, in which case it is often referred to as "mountain-top removal," a true war waged against the mountains. Looking at the whole mining process worldwide, it is estimated that about 10 percent of the primary energy produced today is used for mineral extraction and processing, mainly in the form of fuels, particularly diesel fuel for extraction and transportation.

Today, the great cycle of mining is far from being concluded, but it already shows signs of having run into difficult times, and the production of many mineral commodities appears to be on the verge of decline. Surely we are not running out of any mineral, but extraction is becoming more and more difficult as the easy ores are depleted. More energy is needed to maintain past production rates, and even more is needed to increase them. This conflicts with the ongoing gradual depletion of fossil fuels that provide the necessary energy. Even in the case of fossil fuels, we are not running out, but extraction is becoming more and more expensive.

Is this overarching trend of depletion going to cause a general decline in the mining industry? It is perfectly possible, and we may be going through a century-long cycle that will lead to the disappearance of mining as we know it.

A Roman gold solidus from the time of Emperor Julian, circa 361 CE. It is from the name of this coin that the modern term "soldier" comes, a clear indication of the fundamental role that money played, and still plays, not just in commerce but also in war.

3

Mineral Empires:
Mining and Wars

A glimpse of times so ancient that money just didn't exist (or at least was not used in everyday transactions) comes to us in the Epic of Gilgamesh, one of the oldest works of literature. In this saga we are told that Gilgamesh—the Sumerian king who reigned around 2500 BCE—is a rich and powerful man who owns plenty of gold and silver. But there is no evidence anywhere that precious metals were used as currency in his time. So Gilgamesh's quest was not for gold but for timber, a precious commodity in the largely treeless Mesopotamia of those times. He travels to the region that today we call Lebanon, and once there he kills Humbaba, the custodian of the forest, in order to harvest the trees. Although Humbaba is described as a monster, the saga implies that something was wrong in the way the transaction was carried out, since the story ends with the death of Gilgamesh's best friend, Enkidu, as atonement for the murder of Humbaba. But, without money, Gilgamesh simply had no way to buy timber; he could only steal it.[1]

A much later document dating back to the 11th century BCE tells a similar story, but with a very different ending, since by then the use of precious metals as currency had become commonplace. It is the story of an Egyptian priest of Amon, Wenamon (or Wenamen), who was dispatched to Lebanon to get timber for his temple.[2] Wenamon's saga is rich in adventures and troubles, but it has nothing of the epic traits of the earlier story of Gilgamesh. And Wenamon could do something that Gilgamesh could not have done in his time: he could pay for the timber with gold and silver.

These stories tell us that money was not just a convenient way of trading. The development of currency transformed the world in several steps that eventually led to the present huge financial system, with its complex credit instruments like stocks, derivatives, and futures, and with the accompanying phenomena of boom and bust, financial collapses, bubbles, and the like. So complex is this system that we seem to be losing the capability not only of controlling it, but perhaps even of understanding it. But if we examine the origins of money, we can obtain some hint about how the present system works. We can also see that for a long time money was directly linked to mineral commodities and that, perhaps, it remains more linked to minerals than we may think.

The Birth of Currency

The story of metals as currency brings us back to the earliest times of human history. With the end of the last glaciation, at about 10,000 BCE, agriculture appeared in the fertile valleys of the world—in Mesopotamia, Egypt, India, and China. Among the many changes brought by agriculture were a large increase in population and the birth of cities. The changes were not limited to the number of people and the population density on the land; human social structures were also deeply affected.

Hunting and gathering societies were relatively egalitarian, simply because there was little way to accumulate goods in a nomadic lifestyle. But in agricultural societies land could be owned and agricultural products could be stored. And not just that: horses, cattle, slaves, wives, and more could become someone's property, and property rights could be codified and maintained by a legal, judiciary, and military system. As a consequence society became stratified: some people owned enormous wealth, others owned nothing, and many were reduced to the condition of slaves.

With these civilizations, trade took on a different structure and character. People had always exchanged goods and services, but in earlier times these exchanges were mostly based on the concept of "gift giving." In the relatively simple world of hunting and gathering, there was no need (and no way) to quantify debt. The idea was that people would give when they had an excess and receive when they were in need. But with the increased complexity of the new agrarian civilizations, that method became impractical. At the same time, barter never was a practical way to exchange goods, and so, in complex societies, there arose the need to record commercial transactions and to make sure that credits and debts were paid in due time.[3]

The first such methods developed in early agricultural civilizations were not associated with gold or precious metals but took the form of promissory notes written on clay tablets.[4] A typical clay tablet could record a trade of goods such as cattle, sheep, or grain; it could specify that the debt should be paid at a specific date and could also define any interest to be paid on the debt—in poultry, for instance. The tablet would be broken to pieces when the debt was paid back. Another way of recording debt was the tally (or tally

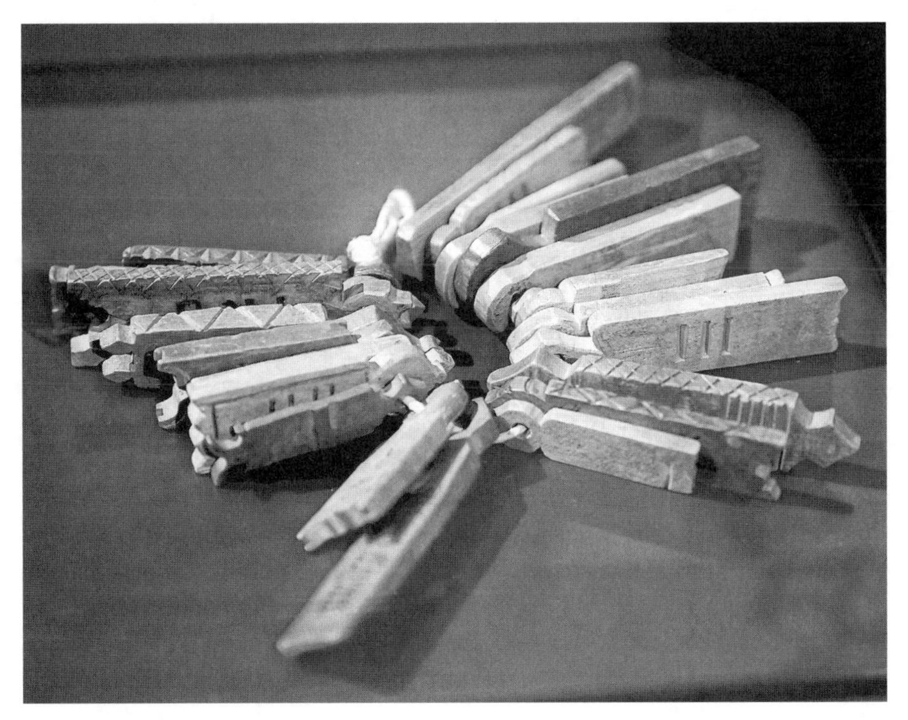

FIGURE 3.1. Tally sticks were still used as way of recording debt and credit in the early 20th century.

sticks)—a notched rod or length of wood that carried the names of the debtor and the creditor and that would be split in half when a deal was struck (see figure 3.1). The two sides of the tally were reunited and destroyed in a fire when the debt was paid. The use of clay tablets disappeared with the 1st millennium BCE, but tallies remained in use in Europe well into the 19th century, and in some cases into the 20th century. Their history is described in detail in an article by Michael Innes, titled "What Is Money," published in 1913 but still well worth reading today.[5]

Tallies can be seen as "money" in all respects. It is likely that they could be traded and exchanged, passing from one person to another many times before they were finally redeemed by reuniting the two pieces. Today there are efforts to create new forms of local currencies,[6] often as part of a movement to build community resilience, such as in the Transition Town and local-living-economy movements.[7] These local currencies can be seen as a new version of ancient promissory notes or tallies and work according to similar principles. But the difficulty in using notes and tallies is in redeeming them in goods and services away from their place of origin. That began to cause problems in early history and eventually led to the development of currency based on precious metals, which has accompanied us up to not long ago.

Metal currency brings us back to the early times of civilization. The specific characteristics of agricultural civilizations made the trade of metals especially important. The fertile alluvial plains that made agriculture possible had formed from the sedimentation of the silt of rivers. In this kind of terrain, easily reachable metal deposits cannot exist. To find exploitable ores, there needs to be the kind of erosion that can be typically found in steep mountain ranges. So agriculture and metal mining don't match, geologically. The consequence was the gradual development of a long-distance trade of metal commodities, as well as the tendency to obtain metals by waging war. This development generated complex societies that didn't base their wealth on agriculture alone. In the long run, the use of metals for trading was also the origin of the great predatory empires of history.

Many early documents and archaeological finds show us that by the second millennium BCE metals were gradually becoming a form of currency. In the Mediterranean region, the code of Hammurabi, dating to about 1772 BCE, provides evidence that gold and silver had become a common means to pay debt and to settle disputes. Similar developments were taking place in China during this period.[8]

With the second half of the 2nd millennium BCE we begin to see a new development: large-scale clashes between different civilizations. In earlier times there had been little incentive for societies to raise up armies and send them beyond their fertile valleys. Think, for instance, of the Egyptians and the Sumerians, two ancient civilizations flourishing at the same time and not very far from each other. There is evidence of reciprocal cultural influence, but none of direct trade or military conflict. Clearly there wasn't much of an incentive to move an army or a caravan across the mountains and deserts that separated Egypt from Mesopotamia. Most likely the Sumerians didn't have much that the Egyptians couldn't manufacture themselves, and vice versa. Besides, most of what could be bought or seized by such an expedition was perishable: grains, sheep, cattle, and even slaves would have been difficult to transport over long distances on land.

But with the diffusion of precious metals, there appeared a good reason for raiding neighbors, even at some distance. As a consequence, we see armies leaving their countries of origin and invading other areas. The very first of these clashes to have been recorded in history was the battle of Megiddo,[9] at around 1460 BCE. It was fought by the Egyptians against the Canaanites who lived in what is today Syria. By our standards it was a minor battle, involving some 10,000 to 20,000 fighters on each side. However, it impressed our ancestors so much that, perhaps, the term Armageddon derives from it. It was the first step toward a kind of warfare that was to revolutionize the world forever.

Some two centuries later, in 1274 BCE, the city of Kadesh, not far from Megiddo, saw another clash of civilizations. Egyptians and Hittites fought there a memorable battle with chariots and infantry that ended, probably, in a draw. We do not have clear evidence that the soldiers fighting at Megiddo and Kadesh were paid in gold or silver, but from what we know of later times, it is at least likely that some form of payment in metal was used to raise and maintain these armies, very large for their times.

While the Hittites were fighting the Egyptians at Kadesh, they left their western border undefended, and there, on the west coast of Anatolia, an army of marauders from overseas sacked and burned a city that the Hittites may have called Wilusa or Truwisa, but that we remember as Troy. In Homer's *Iliad* (written in the ninth or eighth century BCE), we are told that the Trojan War was waged for a woman, and that may tell us something about the reasons for many ancient wars. But the emphasis that Homer himself places in describing the riches of warriors tells us that, already at that time, there were different justifications for military expeditions. Homer tells us explicitly that precious metals, and also iron, were used in transactions. In Homer's *Iliad*, we find gold mentioned 124 times, silver 48 times, bronze 128 times, and iron 48 times. Among other things, we read that King Priam offered 20 talents (half a ton!) of gold to Achilles as ransom for the body of his dead son, Hector.[10] We also read that a block of iron was offered by Achilles as prize at the games that he held for the funeral of his friend Patroclus. The fact that these minerals came from afar is also recorded in the *Iliad* with the mention of the "silver mines of Alybe," although nobody knows what present location corresponds to that name.

These ancient wars were the first symptoms of a deep change in the structure of human society. It was a transition from static agricultural civilizations to aggressive predatory empires, societies that lived mainly on conquest. Of course, peoples have always found reason to fight one another, but imperial conquest involves spending years in campaigns in faraway places—a much larger commitment than a simple seasonal raid. Soldiers for imperial armies might fight out of fealty to their lord or their king or for the promise of booty when the campaign was over. But that doesn't mean that they didn't want to be paid in advance. And payment needs some kind of currency.

Soon precious metals became not just a currency for trade, but a major military weapon that generated a form of enhancing feedback. The more gold a king had, the more retainers he could hire; the more retainers he had, the more gold he could raid from his enemies. As a result, it became fashionable for kings to show off their wealth by appearing clothed in gold and with plenty of gold trappings: crowns, scepters, rings, necklaces, and all the rest. It was, among other things, the beginning of propaganda as an art and a science. One of the earliest examples of such bejeweled kings, dating back to the fourth millennium BCE, can be seen in burial in Varna, Bulgaria (see figure 3.2). We can hardly imagine the aura of power this man would have cast when he was alive.

FIGURE 3.2. The Varna necropolis, found in Bulgaria in the 1970s and dating from 4600 to 4200 BCE. This individual must have been a powerful king or warlord, as shown by the impressive array of gold objects buried with him.

Precious and Noble: How Gold and Silver Supplies Impact the Economy

Luís de Sousa

Are we ever going to run out of gold and silver? Probably not. Almost all the gold that has been mined in the past is still available aboveground in the form of coins, bullion, jewelry, and more. Even though silver has been partly dispersed in nonrecoverable forms, as in electronics and mirrors, large amounts of it, too, remain aboveground and will stay there for a long time if kept as stocks of financial value.

That doesn't mean, though, that there won't be problems with the future availability of these metals in the financial and industrial systems. Both reserves and production data point to short-term mining decline, with different consequences for each metal. Whatever the final outcome, gold and silver will certainly play a role in the definition of the economic paradigm for the 21st century.

To understand why, we first have to understand what has made these metals so important in the global economy.

Gold

Why is gold so precious? It all comes down to four essential characteristics: low concentration in the Earth's crust; even distribution across the crust; chemical stability, which prevents corrosion; and high density.

The low concentration of gold in the crust is often cited as the main reason for its value. Yet, although it is found in small amounts, it is actually present almost everywhere on the planet, a fact that early on rendered it an easily recognizable asset. But the real driver of gold's precious status was its density. Being almost twice as dense as lead and silver—indeed, denser than any other metal known before the 19th century—gold could be made into standard coins that could be easily authenticated with a simple scale. Elements of similar density were not identified until the 19th century, and only one is more abundant: tungsten.

From a monetary perspective, gold's low concentration in the Earth's crust translates to a slow increase of stock. Its even distribution in the Earth's crust makes it a universally recognizable and accepted value. Its chemical stability eliminates intrinsic devaluation. And its density makes it nearly impossible to counterfeit. From early usage as an adornment, gold rapidly became a store of value and eventually evolved into an abstract currency, differing from modern currencies only by its limited supply.

As the most important precious metal, gold had a core role in the monetary policies exercised by institutions throughout history. With the industrial revolution, new energy and commodity flows opened up the way to unprecedented economic growth. The new wealth brought something else that was new: the decadal economic cycle, with regular recessions spreading misery among the new industrial workforce. Though still cause for debate, many credit this cycle to the mismatch between the precious metals supply and economic growth.[11] As the economy expanded, the essentially static supply of gold made it increasingly valuable against industrial goods, eventually leading investors to prefer liquidity in the form of gold to the risk of investment.

The 20th century started with the buildup to an unprecedented confrontation between the industrial nations, in great measure to define access to resources in the rest of the world. During World War I all industrial nations introduced alternative paper currencies to support their industrial effort. In the aftermath of that war they all returned to gold-pegged currencies, probably spurred by the collapse of the deutsche mark. In 1928, industrial activity took a general downturn, and one year later, Black Tuesday brought about the Great Depression.[12] Whatever the role gold may have had in this event, in the 1930s industrial nations were all on track to abandon gold once more. This was to be a slow process, only completed in 1971 when the United States fully depegged its currency.

The emergence of paper currencies provided state institutions with a crucial controlling mechanism over investor expectations. Without any physical links to restrain their supply, paper currencies can be managed so that they never become better investments themselves than tangible assets. In other words, they are abstract, and modern abstract currencies function as stores of value only if properly invested. Without this system, the economic growth of the second half of the 20th century would not have been possible. But for this system to work, central banks have to manage the prices of precious metals. The goal is to avoid the latter becoming more desirable investments than paper currencies.

To this end, central banks built strategic gold stocks, selling or leasing these stocks in order to stabilize prices as necessary. By allowing a tame appreciation, they activate recycling processes that convert jewelry into bullion, thus guaranteeing an influx of metal into the market. The value of gold has been a sort of sword of Damocles over the heads of modern abstract currencies, but so far central banks have managed to maintain control, weathering serious crises in 1968 and 1980.[13] Annual mined volumes of gold doubled from 1.2 ktons (thousand metric tons) in 1984 to over 2.4 ktons in 1998, with a peak set in 1999 at 2.6 kilotons. From then a slow decline followed, until the trend reversed in 2009. In 2010 the volume surpassed 2.7 ktons for the first time and went above 2.8 ktons in 2011 and 2012. Wholesale gold prices increased from $9/gram in 2001 to over $50/gram in 2012.

Geologist Jean Laherrère estimated in 2009 that less than 100 ktons of extractable gold remained to be mined worldwide, postulating that world extraction couldn't go much higher and would soon enter a permanent decline.[14] He noted that the countries that dominated gold extraction in the 20th century—South Africa, the United States, Australia, Russia, Canada, and Brazil—are all well into the decline phase, in tandem with declining ore grades. Conversely, Peru, Ghana, Mexico, Chile, and Uzbekistan are still experiencing growing extraction volumes, but reserves estimates are much lower than for any of the historical producers. Laherrère predicted that production in all these latter countries would peak before 2025.

Today, however, the most important gold-producing country in the world is China, the main force behind the productive trend reversal seen in recent years. China extracted 360 tons in 2011 alone—far more than any other country. Nevertheless, its ultimate recoverable reserves (URR) are much smaller than those of the historical producers; they're estimated by the US Geological Survey (USGS) to be between 8 ktons and 10 ktons, of which around 5.5 ktons has already been extracted. How long the growth gold extraction in China can last is difficult to say, but it is certain that when it stops it will mark the definitive decline of world gold mining.

This coming decline may not bring the gold market under pressure, as the gold mined in the past is still available aboveground. At the end of 2011 the World Gold Council estimated that over 170 ktons of aboveground gold was distributed across jewelry (50 percent), central bank stocks (18 percent), investment assets such as coins and bars (19 percent), and industrial stocks.[15]

This large stock means that the record level of gold extraction in 2011 represented a global stock increase of only 1.6 percent. This number vividly portrays the way gold holds value over time, but it also shows how much less relevant gold mining has become with time. Thirty years ago gold recycling was supplying a little over 300 tons/year to the gold market (around 20 percent of the total supply). In 2009 this figure reached 1.7 kilotons, then over 40 percent of the market. In 2011 the gold traded worldwide amounted to some 4.5 kilotons, an absolute record, but still less than 2.7 percent of the aboveground stock.

It seems the gold market can withstand the coming mining decline, either through direct intervention from central banks, or with controlled price increases to mobilize stocks into recycling. Consider events in the second quarter of 2013, when British investors brought onto the market huge amounts of gold previously held in private stocks. In a matter of weeks the UK became the largest gold exporter in the world, shipping overseas more than 800 tons of the metal—this from a country without a single active mine.[16] Beyond slashing 25 percent off the price of gold, these investors are sure to close down at least half of the world gold mining operations if they keep supplying the market with such volumes.

Silver

Silver is not as valuable as gold, being more abundant in the Earth's crust and less dense. It also tarnishes easily when in contact with air. Its value comes rather from a practical perspective: it is the most conductive and the most optically reflexive metal known.

Nevertheless, silver is sufficiently inert and its supply sufficiently stable that it can be used as money. In the past, in small amounts it was valuable enough to support daily trade but not enough to prompt falsification. Silver became money for the common man, and it took on an important financial role, substituting for gold when lower-value goods were exchanged.[17]

In ancient times the main technological use of silver was for mirrors. In modern times it found many new roles, including the production of photographic film. Its applications continue to expand today in consumer electronics, medical appliances, electric batteries, catalysts, and even clothing. It is also used alloyed with other metals, such as zinc and cadmium.

From the late 1970s up to the mid-1990s the silver supply remained somewhat stable, never surpassing 20 ktons/year.[18] During this time mined volumes went up by a third, with recycling declining in equal measure. Between 1994 and 2001 the total silver supply grew by more than 40 percent, up to 27 ktons/year, according to the Silver Institute.[19] The rising trend continued throughout the next decade, though at a slower pace, up to a record of 32 ktons in 2011. Recycling supplied a fairly stable share of the market, meeting around 23 percent of demand during that decade. Mining grew in share, from 64 percent in 2002 to 77 percent in 2009 and 73 percent in 2011, making up for a decline in industrial stock drawdowns.

Since 2000 the industrial use of silver has remained remarkably stable, averaging between 17 and 18 ktons/year. Consumption for photography purposes decreased from 6.3 ktons in 2002 to 2 ktons in 2011, due to the rise of digital equipment, but other industrial applications grew in equal measure. Overall, silver demand has expanded mostly as the result of speculative investment, which went from virtually zero in 2002 to 5 ktons in 2011.

This speculative demand reflects a reaction by investors anticipating short-term supply constraints. Recent USGS estimates point to remaining reserves of 500 ktons, equal to only 15 years of supply (or 20 years of mining).[20] However, the same institution was issuing estimates below 400 ktons as late as 2005, and some observers have projected even lower supplies.[21]

This perceived scarcity is accompanied by what still is a historically low silver price. Up to the 19th century the value of silver remained basically stable relative to that of gold. In Roman times the ratio of gold to silver in value was 12 to 1, meaning that 1 gram of gold was worth the same as 12 grams of silver.[22] By the late 18th century governments were setting the value at 15 to 1.[23] These ratios largely reflect the relative abundance of these two metals in the Earth's crust: for each gram of gold in the crust there are about 18 of silver.[24] After 1900 silver progressively lost value against gold, reaching a low of 100 to 1 in 1990 and hovering around 55 to 1 today. This devaluation of silver is possibly associated with modern mining techniques, whereby silver is obtained through catalytic refining of ores extracted in mines dedicated to other metals like copper, nickel, and zinc.

This depressed price has promoted the loss of silver stocks. Silver dispersed in cheap jewelry, outdated coins, photographic film, obsolete electronic devices, and other items has been ending up in dumps, and some of it might have even already been lost at sea (in the form of finely dispersed particles eroded from silver artifacts), from where it will never be recovered. The result is a relatively small industrial stock of silver, equaling about 25 kilotons—less than 4 grams per person on the planet, less than one year of mining supply,[25] and less than one-sixth of the world's gold stocks.

The low prices of silver also have had an impact on non-industrial stocks. However, there is no official accounting of this material, so it's impossible to estimate those stocks with any accuracy. A recent attempt by analyst David Zurbuchen placed these stocks at 650 ktons.[26] Even if this figure is accurate, it represents only 20 years of supply. For comparison, the world gold stock is equal to more than 60 years of supply.

All this makes for an unsustainable scenario in the coming years: growing demand, dwindling reserves, uncertain stocks, and prices unaligned with physical abundance. This scenario could lead to three outcomes:

- an increase in silver recycling, with a relevant rise of nonindustrial stocks flowing to the market;
- the evolution of mining toward silver-dedicated mines, if lower ore grades are technically feasible; and
- the substitution of silver by copper in industrial applications where possible.

All of these outcomes, not mutually exclusive, will certainly require considerably higher silver prices, and possibly a return to the historical silver-to-gold ratio. This poses a serious challenge to central banks, which largely lack mechanisms to fight liquidity runs into silver.

The Outlook

Gold and silver are not precious by chance, and considering that two-thirds of gold and three-fourths of silver reserves have already been mined, they will certainly retain their value in coming years.

Constrained access to other commodities, especially energy, and the over-indebtedness of states and citizens could paint a dire scenario for investors. If governments and central banks opt to restrain paper currencies, they also restrain industrial demand for precious metals. In the process they will face serious social consequences that may not be sustainable in the long run. If they choose to loosen monetary policies, the relative value of scarce commodities, silver in particular, can cause uncontrolled price rises.

Coinage as a Military Weapon

The use of precious metals as means of commercial exchange—in other words, currency—generated a tremendous increase of both commerce and warfare. In time, it led to the appearance of the first military empires in the Middle East, and those empires expanded to cover large swathes of territory that were kept under the rule of the central government by military means. At the same time, a widespread network of commercial activities started to appear, especially in the Mediterranean region, where navigation provided an easy and practical way to carry goods. At the beginning of this historical phase precious metals were exchanged in the form of bullion, but bullion necessitated a laborious process of weighing whenever a transaction was made. A more portable and efficient method was needed in order to bring metal currency to the hands of every soldier and trader. That turned out to be coinage—a technology that is believed to date back to the middle of the first millennium BCE. It is possible that the Chinese had already invented coins by the 10th century BCE, but in the Mediterranean region the first coins were minted in Lydia, in western Anatolia, at around 550 BCE.

Coinage was a remarkable feat of metallurgy for the time as it required molds, or dies, to impress an image on one or both sides of a silver or gold disk.

FIGURE 3.3. *Top left:* The two sides of a Lydian coin in electrum (a gold-silver alloy), dating from the sixth century BCE. Note how the coin is one-sided, formed by striking a hammer against a die. This characteristic was maintained by all coins of the Persian Empire. *Bottom left:* Silver tetradrachma from Athens, circa 450 BCE. It shows the classic symbol of the goddess Athena, the owl. Note that it is two-sided, struck on both sides by two different dies. *Right:* A Persian "daric," circa 420 BCE. This coin shows its derivation from the Lydian technology because it is one-sided.

These dies had to be very hard if they were to be able to be struck hundreds or thousands of times against gold and silver disks and still maintain their ability to leave an impression. That, in turn, created the problem of what tools to use to engrave the die. Making good dies for coinage required highly skilled craftsmen and advanced technologies for the time. The ancient dies that come to us from the archaeological record are made of bronze or iron. They were likely engraved before being surface-hardened by the same kind of methods used to make steel for sword blades.

As it often happens, rapid technological advances arrive in times of great need. By the sixth century BCE the Persian Empire (more exactly, the Achaemenid Empire) was growing in the Middle East by gobbling up its neighbors, one by one. On the path of its expansion westward, the Persians confronted the Lydian kingdom in Anatolia. The Lydians seem to have put up a spirited resistance, but eventually they were overwhelmed and absorbed. It is at about this time that the last Lydian king, Croesus, is said to have invented coins (and given birth to the saying "as rich as Croesus," which is still known today).

The ancient reports on Croesus's invention are paralleled by archaeological evidence. Disk-shaped objects that we could call "coins" were found in modern archaeological excavations in the area corresponding to the ancient Lydian kingdom, in what is modern Turkey. These coins were made of the silver-gold alloy electrum and carried the effigy of a lion. Their standardized weight made it easy to distribute them: coins could be simply counted and didn't need to be weighed. Each one of these metal disks was the embodiment of a credit to the owner from the king who had issued it. But unlike the old promissory notes written on clay tablets or tallies, the coin was only one object, held by the creditor. It had no counterpart with the debtor. Using coins, the debtor would not lose his properties or be enslaved because he couldn't honor a contract he had signed on a tally. But what guaranteed the creditor that his "half tally" could be redeemed? The value of gold was known and accepted everywhere, so the creditor could be sure that the coin could be redeemed everywhere—even far away from where it had been issued and even if the king who had issued it was defeated or died. At worst, the coins could be melted down and new coins could be minted with the mark of the new king on them.

The rarity of gold and silver also guaranteed that the number of gold coins a king could mint was limited by his ability to steal gold from other kings, have his subjects pay taxes in gold, or conquer and control gold mines. The invention of coins, of course, raised the problem of falsification—another practice that needed sophisticated technologies. Even in very early times kings tended to trick their constituents by giving them coins not made of pure gold but alloyed with copper, silver, and other less valuable elements. Since alloys tend to be harder than pure metals, people soon discovered that a good way to test whether a coin was pure gold was to bite it. If teeth could leave an impression on the metal, then it was most likely pure gold. The color of the coin could also be an indication of its purity, which led to the development of "touchstones," where coins would be rubbed to gauge the color of the impression they left. A coin's weight could also be subtly reduced, for instance by filing its edges. (That old trick is why the edges of modern coins are reeded. If any of the precious metal is filed away, it would be easily noticed.) An even more sophisticated technology consisted in plating the surface of a copper (or other nonnoble metal) coin with a thin layer of gold and silver.

Money counterfeiting seemed to carry a peculiar fascination in ancient times just as it does today. But it is a difficult and expensive activity that also carries big risks and harsh legal punishments. So the development of counterfeiting technologies didn't prevent the Lydian invention of coinage from being a huge success. The Persians also rapidly adopted the technology of coinage, perhaps taking back home the same craftsmen who had worked for the defeated Lydian king Croesus. The coins minted by the Persian Empire clearly show their derivation from the Lydian ones in being "one sided"—that is, they were struck with a hammer against a single die.

In the same period the Greeks developed their own, more advanced coin technology. The Greek drachma, most often in silver, was struck between two different dies to emboss an image on both sides. The struggle that took place between the Greek city-states and the Persian Empire could be seen, in many respects, as a fight between two currencies: the daric on one side, the drachma on the other. After the defeat of the Persians at the battle of Salamis, it was this Greek coin that dominated Mediterranean trade for centuries. In general, coins carried symbols of the kingdoms and cities that had coined them but were exchangeable with similar coins with the same weight. The situation was not unlike that of the euro today in Europe, where each nation mints coins with different symbols, but all are interchangeable with each other.

Mineral Empires

If precious metals made empires, where exactly did those metals come from? It is tempting to assume that the control of gold mines drove the expansion of most major historical empires. Unfortunately, we usually don't have quantitative data on the yield of ancient mines—only very uncertain estimates. So it is impossible to know for sure how much gold each empire produced itself, and how that correlated to political and military power. There are enough hints, though, to suggest the correlation was strong.

Gold from alluvial deposits is relatively common in many regions of the world, or at least it was common in ancient times, before it was extensively mined.[27] It made its way into riverbeds as rain washed over ore deposits, usually composed of gold dispersed in quartz, and carried nuggets downstream, where they could be inexpensively panned. Alluvial deposits, however, can be rapidly exhausted, and mining must then move to the origin of the nuggets—the ores. Mining ores is much more difficult and requires hard work and considerable investments. Here we see another case of reinforcing feedback created by mineral resources: kingdoms that had gold could invest it to pay (or enslave) miners to extract even more gold.

Most ancient agricultural civilizations had access to at least some gold, as the archaeological record and ancient documents show. The Nile was too slow to carry nuggets to placer deposits, but it is known that the Egyptians mined gold veins located in Egypt's Eastern Desert. The same was true for the Tigris and the Euphrates rivers in Mesopotamia; they weren't swift enough to carry gold downstream. So the Sumerians obtained their gold, probably, from the mines of Zarshuran, in the region that is today Iran.[28] It may be that the

Egyptian gold mines never were productive enough to propel Egypt to the status of world power, but those of Zarshuran may have been the origin of the Persian Empire, one of the largest that history has ever seen, and of the power struggle for domination in the Mediterranean region that started with the second half of the first millennium BCE.

In the sixth century BCE, the Persian Empire managed to control most of the metal resources of the eastern Mediterranean and Middle East. But on the western edge of the Persian Empire the city of Athens, beyond the western border of the empire, managed to control the silver mines of Laurium, southeast of the city. Laurium was the site of one of the richest deposits of precious metals of that age, and its mines played a fundamental role in the conflict that pitted a coalition of Greek cities against the Persian Empire led by King Xerxes. Athens used the revenues of the Laurium mines to build the powerful military fleet that destroyed the Persian fleet at the Battle of Salamis in 480 BCE, putting an end forever to Persia's attempt to expand into Greece. Empires are by their very nature unstable structures; they can exist only by either expanding or contracting. With the defeat at Salamis, the Persian Empire entered an irreversible spiral of decline, perhaps also caused by the depletion of its gold mines.

Instead, the silver of Laurium pushed Athens to a brief imperial period in which it dominated the central Mediterranean region. Athens declined with the decline of the Laurium mines, while the rise of the Macedonian kingdom, with Philip II, seems to have been linked to the discovery of silver in Macedonia and to development of mining there.[29] It may have been because of these silver resources that Philip managed to conquer Greece, succeeding where the Persian king Xerxes had failed. Later, Philip's son, Alexander "the Great," went on to conquer Persia and to create a vast empire that reached up to India. The decline of Alexander's empire may be related to the decline of the Macedonian silver mines that had produced it. In time the lead passed to the western Mediterranean region, which still had largely untouched mineral resources.

Rome got its start as a small agricultural village in central Italy. There was no gold in the immediate vicinity, but as they expanded, the Romans took control of the Tuscan copper mines and used them for their coins. The economy of the early Roman Republic was based on copper and bronze rather than gold, and that gave the Romans a reputation for being frugal and tough warriors. Soon, however, the expansion of the republic led Rome to conquer gold-producing regions in the Italian Alps and Sardinia. At this point the power game in the western Mediterranean became a conflict between Rome and

Carthage, a North African city located in what today we call Tunisia. Carthage started as a Phoenician colony, but it had rapidly grown to the status of an imperial city that mined gold and silver mainly from Spain. The struggle between Rome and Carthage lasted more than a century and ended with the destruction of Carthage in 146 BCE. After that, the Romans had a free hand to exploit the mines of Spain.

The abundance of gold and silver in Spain may have been the element that propelled Rome to domination over the whole Mediterranean region and most of western Europe. The last phase of the Roman expansion in Spain came in the first century BCE with the conquest of the northwestern regions that we call Asturias and León. Soon these regions would become the largest source of gold and silver in Europe for a few centuries. The control of these mines gave to the Romans a wealth that had never been seen before in Europe.

The Roman society was a structure dedicated to war, its main economic activity. In this sense the Romans used money largely as a military technology. With money, they paid a standing army, one of the first recorded in history. They also used money to pay auxiliary troops that augmented the Roman legions. Finally, they used money to bribe enemies. Especially during the last period of the empire, it was common for Romans to buy off enemies rather than fight them. The mechanism worked wonders, at least for as long as the Romans had gold and silver available to them.

The Roman approach to war was that of a commercial enterprise; it had to create a profit. So the Romans did very well against societies that were similar to their own but outmatched in terms of military resources. In conquering the Hellenistic states and Gaul, they could bring home booty in terms of precious metal and slaves that repaid their expenses for the campaign and allowed them to start new ones. They fared much less well against enemies, such as the Scots and Germans, who didn't use metal coins and were too poor to provide a sufficient booty to justify a campaign.

In time the Romans started to face big problems of monetary supply. We don't have data for the production of the Spanish mines in ancient times, but we know that from about 50 CE the Roman denarius started to contain less and less silver. By 250 CE it was pure copper. It is very likely that the debasement of the denarius resulted from depletion of the Spanish mines. The Roman gold coin, the aureus (and the later solidus), didn't go through the same process of debasement, but it is likely that smaller and smaller numbers of coins were minted as the production from Spanish mines declined.

Apart from gold and silver, the Roman Empire never produced much more than two things: legions and grain, neither of which was a tradable commodity with the outside world. So the Romans imported all sorts of luxury products from Asia and the Middle East: silk, spices, ivory, pearls, slaves, and more. They paid in gold, and that gold never came back because the Romans had little that they could sell outside their borders. Gold and silver also disappeared from the empire as foreign mercenaries took their pay with them when they went back home. And in the last period of the empire, a perverse negative mechanism took place: deflation. With gold becoming rare, it became more and more valuable, so people tended to hoard it. Many buried it underground, removing it from circulation in the economic system. That buried gold was perhaps the origin of the medieval European legends of dragons hoarding gold in their lairs.

With the second century CE, the Roman Empire attempted its last feats of conquest. Under Emperor Trajan, it managed to annex Dacia, a kingdom located in the region of central Europe that corresponds to modern Romania and that controlled gold mines in the Carpathian mountains. Then, perhaps with the use of the gold looted in Dacia, Trajan attempted a thrust into Persia and Arabia. The idea was, probably, to recover some of the gold that the empire had lost through trade, alongside taking control of the caravan routes to Asia. It was the last of several Roman attempts to conquer, or at least control, the region. It ended in failure. Asia was too big for the Romans to conquer, while Arabia was too dry and too hot.

With the failure to recover its lost gold and silver, the Roman Empire was doomed, at least in its western half, which had run out of mineral resources that could be extracted with the technologies of the time. By the fifth century CE, the last century of the Western Roman Empire, coinage had basically disappeared in Europe, except in forms that seem not to have been used as currency, such as medallions or decorative objects. There are reports that Roman soldiers were paid in pottery, and the military paradigm of the last centuries of the Western Empire was that of the *bucellarii*—literally "biscuit eaters," people who fought for their masters in exchange for food. Soldiers were also paid with parcels of land (which led, in part, to the feudal system that was to replace the empire in Europe).

During the Middle Ages, the Eastern Roman Empire never regained the military power that had been the characteristic of the empire as a whole, but it maintained a tight grip on some of the most profitable caravan routes in the Middle East. It continued to mint the gold solidus or (in Greek) nomisma, which remained the standard and most diffuse gold coin in the region up to the 11th century. These coins were minted probably using gold traded by the neighboring Asiatic and Arabic countries. In western Europe, though, the dearth of precious metals continued. Gold was so rare that it often appeared in the form of "bracteates," coinlike decorative objects so thin that they could be engraved only on one side, the back showing a negative image of the front. They were minted by striking a pile of thin metal disks against a leather stand.

In the time of Charlemagne (742–814) the dearth of precious metals in Europe seems to have become less serious. New silver mines were discovered in eastern and northwestern Europe, for instance the mine of Rammelsberg, in Germany.[30] So Charlemagne adopted a pure silver standard as part of a minor European renaissance during his reign. Later, more new silver mines were discovered, such as in Freiburg.[31] These mines may have been an important factor in the economic growth of late medieval Europe.

While the Europeans were busy with their feuds, the Arabs put to good use the gold that they had gained in their trade with the Roman Empire. They embarked on a campaign of conquest that led them to create a new empire embracing North Africa, Spain, and most of the Middle East. With the dynasty of the Umayyads, the Arab caliphate reached its greatest extension during the seventh and eighth centuries. Afterward its expansion ceased. Like all empires, the caliphate could not survive without expanding, and it started its trajectory of decline. In the meantime, many things were changing in the world.

Global Commercial Empires

During the Middle Ages, with southern and eastern Europe badly depleted in mineral resources, there was no way to rebuild empires based on gold, as the Roman one had been. Yet the end of the Middle Ages was a period of rapid economic growth for all of Europe, and in particular for Italy, where the Renaissance had originated, with the rapid rise of local powers, such as the seafaring republics (Amalfi, Pisa, Genoa, and Venice) and industrial and commercial cities such as Florence. Perhaps for the first time in history major world powers were based not on military might but on commercial power. The European industries, mainly the Italian textile industry, were reaping huge profits by exporting their products to the East, in large part to the economically declining Middle East and Northern Africa. There were no gold mines near Florence or in areas it controlled, yet its merchants were able to bring in enough gold that Florence started minting its own gold florin in the 13th century, making a strong departure from the Carolingian silver standard (see figure 3.4). To this day Florentines take their oath by swearing on the image of St. John that the florin used to carry.

Copper: The Near-Peak Workhorse

Rui Namorado Rosa

No metal other than copper has the same combination of low price, high electrical and thermal conductivity, good resistance to corrosion, and good mechanical properties, especially when alloyed with other elements. Indeed, copper has been extensively used in human history. Over the centuries, we have extracted and dispersed enormous amounts of copper, and production continues to increase, reaching today the highest levels ever. Copper was one of the first metals ever extracted and smelted, and most likely it was the first metal ever used to produce cutting implements, weapons, statuary, and other items demanding a strong, resilient material. With the industrial revolution, copper and copper alloys were used for machine parts, taking advantage of their strength and resistance to corrosion. Over time, many of copper's uses as a structural material were taken over by steel, which could be better protected from corrosion. Still today, however, the good resistance of copper and copper alloys to atmospheric corrosion makes them useful in a variety of circumstances where long-term performance is needed.

Today, copper is alloyed with tin to form bronze and with zinc to form brass. It is also alloyed with aluminum and other elements to form light alloys often used in the aerospace industry. The main use of copper today, though, derives from its excellent electric conductivity, which is second only to that of silver. Copper is vital for everything that has to do with transporting electrical current, from transmission lines to electrical motors. It is also fundamental in the electronics industry, where it is used in a variety of applications. The average per-capita stock of each inhabitant of the developed world is reported as about 140 to 300 kilograms of copper, an indication of the great importance of this metal.[32]

But as for all mineral resources, copper ores exist in limited amounts. Copper's production has increased exponentially during the past 40 years, at an average yearly rate of 2.3 percent—a rapid and sustained growth that has kept in check, so far, all fears of a possible decline generated by depletion. However, there are worrisome signs that not all is well in the copper world.[33] A number of facts hint at an imminent slowdown of copper availability:

- Accumulated past production will exceed remaining reserves in one decade.
- The extraction rate has exceeded the discovery rate for two decades.
- Only 56 new important discoveries of copper deposits have been made in the past 30 years.
- Only 7 of the 28 largest copper mines in the world are thought to be amenable to expansion.
- Many large copper mines will be exhausted in the coming years.

While production has been growing, the grade of the minerals mined has been steadily declining. As a consequence, mining becomes more and more expensive. At the same time, while the search for new resources has led to a remarkable growth of the known reserves, the reserves-to-production (R/P) ratio has remained close to 30 years of supply.[34] So we need to ask some basic questions: For how long can the present growth can be sustained? And for how long can the present levels of production be maintained?

The total amount of extractable copper of all types in the continental crust is estimated at nearly 1,800 million metric tons.[35] Potential resources of copper could also be found on the deep-sea floor in the form of dispersed manganese nodules and crusts. The total mass of the seafloor resources is very large, estimated at 13 billion metric tons.[36] But extraction is extremely costly and the copper content is just a few percent.

The discovery rate of new copper resources has stayed close to 7 million metric tons per year for the past 150 years.[37] This figure should be compared to the rate of primary production of copper, which was about 6 million metric tons in 1970 and 9 million metric tons in 1990, and recently attained 16 million metric tons per annum, meaning that the world production has exceeded reserve growth for the last two decades.

The United States was gifted with a large endowment of copper and was home to several of the technological innovations in copper mining and beneficiation (the process of separating the targeted mineral from extracted ores). US Geological Survey reports show that the amount of copper extractable from ore, though, has gradually declined. Mined ores contained about 10 to 20 percent copper around 1850 but had dropped to 3 to 4 percent in 1900, and by 1970 the average ore grade had dropped to 0.5 percent copper. In the past few decades, there has been a major shift to the practice of heap-leach mining, using solvent extraction and electrowinning (techniques that use chemicals or electric power to extract the copper) to capture lower-grade porphyry ore (0.2 to 0.5 percent copper). The remaining reserves in the United States are smaller by far than cumulative production, indicating that peak production, observed in 1996–97, is irreversible. The United States was once a major producer of copper, but today it has become a heavy importer.

Canada, which was among the top five copper-producing countries until four decades ago, passed its peak production of copper in 1973, while discoveries had peaked around 1965. Exploration efforts led to important new discoveries in the 1960s, mostly in porphyry ores, and the reported reserves peaked in 1983. Annual production trends over the past few decades indicate that the country's R/P ratio has already declined to 13 years.[38]

Copper mining in Australia was based on ore grades of 15 to 25 percent copper content before 1880, but that percentage rapidly declined, to about 5 percent in 1900 and just a few percent during the past two decades. Since 1950, Australia has consistently reported growth in economic copper resources, due to reserve growth at known mines as well as the discovery of major new deposits. Australian copper reserves total 79.6 million metric tons at 0.86 percent average grade. Copper in some proposed projects is estimated at 15.96 million metric tons at 0.46 percent average grade. As a consequence of declining ore grade, the specific water and energy consumption has risen rapidly.[39]

Chile holds a quarter of the world's acknowledged copper reserves and is by far the largest copper producer in the world, contributing about 35 percent of the total. The country's total output has been nearly flat since 2004, however, a pattern that suggests existing mines are approaching maximum production capacity. The country still has vast copper reserves, but their average ore grade declines continuously.[40]

It appears that the United States, Canada, Zambia, Zaire, and most of the small producers have already exploited more than half of their resources, indicating that peak production has been reached. Countries in the former Soviet Union (Russia, Armenia, and Kazakhstan) appear to be close to peak production, as reserves and cumulative production are almost in balance. China and Indonesia appear to be close to maximum capacity as well, given an R/P ratio close to 30 years. Chile, Peru, Australia, Mexico, and Poland all seem to be still well behind peak production.

If we consider that the total world endowment of extractable copper is estimated at about 1,800 million metric tons, and that a total of 600 million metric tons have been extracted from the primordial endowment so far, the remaining resources could seem to be abundant.[41] But the problem is not the amount but the cost of extraction, which has been increasing due to the progressively diminishing grade of the resources being processed.

As ore grade declines, the volume of ore mass that must be extracted and hauled to obtain a unit of copper grows. That means vast amounts of energy are required for copper mining. In addition, lower ore grade requires finer grinding and milling to free the smaller proportion of copper-bearing minerals. These two factors increase drastically the energy demand per unit of product.

In Chile, for instance, the energy required to mine copper rose by 50 percent from 2001 to 2010, but the total copper output increased just 14 percent.[42] The associated increase in electrical energy demand for the whole copper mining sector is forecasted to grow at 6 percent annually—that is, at a faster rate than material throughput. The US copper mining industry has also been energy hungry.[43] The energy intensity of copper recovered at the mine gate in the United States is four times larger than the figure reported for Chile.

The declining quality of raw materials and rising energy and material costs in primary copper production are strong incentives for resorting to more systematic metal recycling policies. However, world recycling of end-of-life copper (old scrap) accounts for only 17.5 percent of total annual copper consumption.[44] The main reason for the relatively low availability of old scrap is that copper products have lifetimes of at least 10 and over 45 years. Most of the copper that has been extracted in the past few decades is still in use, and more is being added yearly to the global economic stock than is being discarded.

So what is the outlook? The signs that Chile, which produces one-third of the world's copper, may move into irreversible decline suggest that Chile's output will plateau and the world's copper output will peak soon afterward. Indeed, some studies suggest the possibility of copper production peaking in a medium-term future, around 2023. When total copper production, including recycling, is modeled into the projections, the peak is postponed to about 2040, with production falling off thereafter.[45]

Either way, the decline of primary copper production is impending, and only a serious rethinking of the way we use this fundamental resource will avoid shortages and the crippling of an important sector of the world's industrial system.

But Florence and the other Italian seafaring republics were limited in their imperial ambitions by their geographical location, which confined them mainly to the Mediterranean Sea. So the expansion toward newly discovered continents soon passed to the hands of the western European states, initially mainly Spain and Portugal, then Britain. The expansion of these new powers came with the development of a weapon that had no equal in history before that time: cannon-armed galleons.[46]

Galleon ships were a remarkable innovation compared with the old oar-powered galleys that were mainly used in the Mediterranean Sea. The need to feed rowers enormously limited the range of galleys, whereas galleons, powered only by sails, could navigate for months at a time. But what made galleons a fearful military weapon was the cannon, itself the result of technological advances in metallurgy and mining. The black powder that made the cannon able to fire was the same substance that allowed Europeans to break up rock to mine enough iron and bronze to make the massive cannon.

Artillery was a remarkably complex technology for those times. It required considerable skill and large amounts of metal to create weapons that wouldn't blow up when fired, a rather common event in those early times. Up until the 19th century, bronze cannons were considered better and more reliable than iron ones, but bronze was much more expensive. In both cases, massive

FIGURE 3.4. A gold florin struck by the Republic of Florence in 1462. Florence was not a strong military power and controlled no gold mines. It was commerce and industry that brought gold to the town. This coin shows the inscription "S. Iohannes B"—that is, St. John the Baptist.

amounts of charcoal were needed to smelt the metal and cast it, and that put the European forests under heavy pressure.

Northern Europe had emerged from the Middle Ages with relatively intact forests, but that was not the case for the old Turkish Empire, which, because of its drier climate, had much smaller forests, if any, and had troubles obtaining enough charcoal to smelt iron. As a consequence, Turkish armies and fleets were hopelessly outmatched by the artillery of their European adversaries. Already, at the battle of Lepanto in 1571, the Turkish fleet had been defeated mainly because of European superiority with firearms. For centuries afterward the Turks remained stuck with their old galleys, and when venturing to arm them with artillery they would uselessly waste resources and money casting monster pieces that were more for display than for effective use. On the other side of Eurasia, the Chinese had oceangoing ships almost equivalent to the European galleons, but they never succeeded in arming them with heavy artillery. It would take time for Europeans to develop portable firearms that would give them a decisive advantage on land, too, but their naval superiority was sufficient to give them the domain of the world's seas. With that, they would start on their path to global domination.

The buildup of the global empires that started with the Portuguese and Spanish was based on the availability of a combination of resources, not just gold and silver. With precious metals troops could be paid, but troops needed firearms to be effective. To make firearms metals were needed, but also wood to make ships and the charcoal needed to smelt and cast metals. "No wood, no kingdome," said Arthur Standish in *The Commons Complaint* of 1611. So the management of forests became a crucial strategic priority for the new maritime powers. But if states needed wood and iron for their warships and their weapons, they also needed food for their troops and their population. For that, it was necessary to clear as much land as possible for agriculture. It was a difficult strategic choice: how to keep a country's forests and at the same time feed its population?

Eventually these mutually incompatible needs put a halt to the expansion of the Portuguese and Spanish empires, even though both had plenty of gold to pay their troops. Neither Spain nor Portugal had enough fertile land to feed its population and, at the same time, grow the forests needed to obtain enough wood to support the military needs of a world empire. The struggle for world domination was won when Britain played a trump card in the game: coal.

FIGURE 3.5. This painting, from around 1666, shows cannon-armed Dutch warships. The emphasis that the painter placed on the display of cannons indicates their importance in naval warfare.

Fossil Empires

Starting in the 18th century Britain became the first empire in the world to base its wealth on fossil fuels. With its abundant coal resources, Britain could produce plenty of iron for cannons. With her powerfully armed fleet, Britain could get timber from anywhere in the world without needing to overexploit her forests. More timber meant more warships, and more warships meant more world domination and, therefore, even more timber. Weapons and warships also meant that powerful armies could be ferried overseas. Everywhere in the world Britain conquered foreign kingdoms and transformed them into colonial plantations that produced food for their remote rulers. More food meant larger armies, and that, in turn, meant more plantations and even more food. It was this self-reinforcing mechanism that created the British empire, the first global empire in history. At the height of national coal production, in the 1920s, the coal produced in England could have matched the heat produced by burning almost all of the world's forests.[47]

From the 18th to the mid-20th century coal was the basic tool of world strategic domination. Other coal-producing nations tried to match the English push but never could really compete. France was producing coal even before Britain, but French coal production never reached the same volumes of the British production. In 1816, when French coal and British coal clashed at Waterloo, British coal won. That was the end of all French ambitions to build a world empire. Germany was slower than both France and Britain in developing coal resources, but in time it did create a powerful industrial economy using the abundant coal resources of the Ruhr region. In the 20th century, Germany arrived to levels of production that nearly matched the British ones. But coal-based empires tended to clash against each other, and in 1914 German coal started to battle British coal. Once more British coal won, this time with some help from American coal, which had been growing too. (See figure 3.6.)

The strategic importance of coal wasn't always limited by the availability of coal mines. Wherever coal could be transported it created the conditions for industry to develop. For instance, northern Italy had good waterways and could develop a local industry using British coal imported by colliers, sailing ships dedicated to coal transportation. On the other end of the Italian peninsula, the drier southern Italy lacked waterways and couldn't industrialize as fast. By the mid-19th century northern Italy had become rich enough with British coal that it could annex the south after a short military campaign. A lack of waterways also plagued the North African and Middle Eastern countries, preventing their industrialization and making them easy prey for European

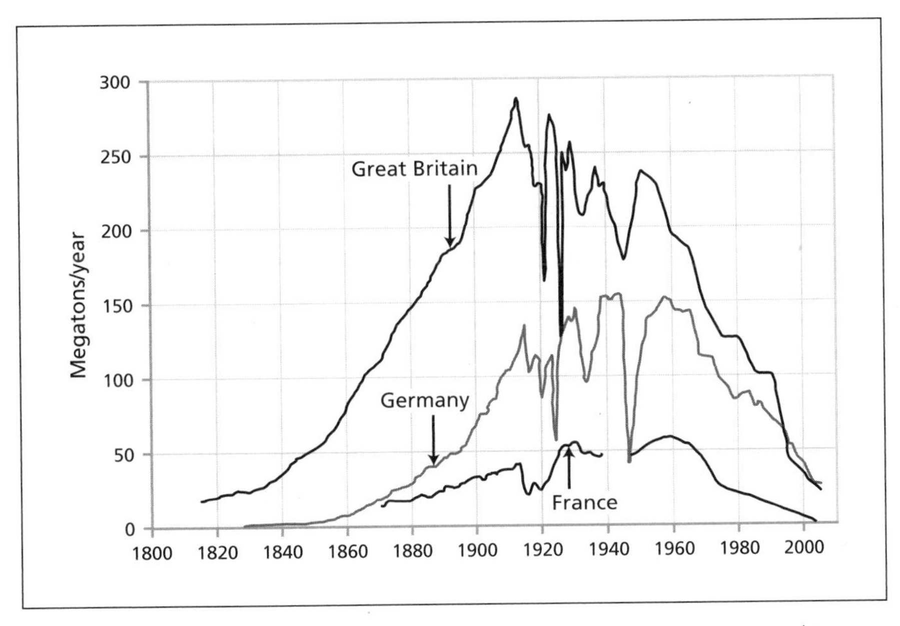

FIGURE 3.6. Coal production in the most important European producers: England, France, and Germany.

powers. The process started with Algeria, conquered by the French from 1830 to 1847. The domination was completed with the First World War, and by 1918 all the North African and Middle Eastern countries were under European control in one form or another.

But coal was soon to start its terminal decline, dooming the coal empires. The coal production curve of European powers tells us a lot about the destinies of their overseas empires. With the gradual decline of coal production, political power also waned, and so ended the British empire—the widest and most powerful empire that the world had seen up to then. British coal production peaked around 1917 and then started its irreversible decline.

Even before coal began to decline, the world saw the arrival of a competitor: crude oil. The clash of these two fuels began with the evolution of naval warfare. In the 19th century, the military fleets of the world had been dominated by ironclads, heavily armored ships powered by steam engines. With time, these warships became progressively larger and armed with more powerful artillery. By the turn of the century, it had become clear that the steam engine was too heavy and not powerful enough to propel this new generation of warships, to say nothing of the vulnerability of the large amount of coal that had to be stored on board. In 1905, the Battle of Tsushima, off the coast of Japan, showed exactly how vulnerable these slow ships were when a modern Japanese fleet wiped out a Russian fleet of older battleships that had laboriously steamed there all the way from Europe. In 1906, the *Dreadnought* battleship was launched in England. It was the first of a new class of battleships that would bear its name for decades. The *Dreadnought* was propelled by a steam turbine that could be powered by different fuels. Of these, crude oil provided the highest power for the same weight and volume stored on board. That made the dreadnought the battleship that dominated the sea for at least three decades. From that time on, crude oil became a strategic resource, and much of the military history of the world reads as the attempt of world powers to secure for themselves the oil resources they needed for war.

The Second World War was, in many respects, a war for oil. Dreadnoughts were already out of fashion by that time, but fuels derived from oil powered all kinds of other weapons: planes, tanks, submarines, carriers, and everything else that moved on the battlefield. A decisive factor in the war in Europe was control of the oil resources in the Caspian area. The German push to Stalingrad aimed at obtaining these resources and keeping them out of Soviet control. The clash was especially bloody, with the number of casualties variously estimated by historians, but always over one million. Defeated at Stalingrad, the Germans refused to quit and kept fighting using synthetic gasoline manufactured from coal. But the final result of the war was another demonstration of how King Coal had been dethroned by crude oil. In Asia the Japanese had initially succeeded in securing the Indonesian oil resources, but in practice they had no hope against the oil giant that the United States had become. In the end it was American oil that won the war.

The Second World War left a world divided in two, and each half based its power on its initially abundant oil resources. On one side stood the United States and its allies, on the other the forces of the Soviet bloc. The competition between these two modern empires never took the form of open warfare, and for almost half a century the two sides faced each other in the Cold War, waged mainly by propaganda. In the meantime, atomic weapons had been developed and both sides were soon endowed with sufficient nuclear power to be able to destroy each other several times over. But despite enjoying the name of "strategic weapon," the nuclear bomb never had a real strategic value since neither side could develop a usable strategy to obtain a military advantage from their possession of it.

If open warfare was never a strategic option in the Cold War conflict, that doesn't mean that there was no struggle. Both sides tried to gain the upper hand by developing a growing economy that would eventually overcome the other in terms of industrial and technological output. That effort was extremely costly in terms of resources, particularly mineral resources. Both sides were well endowed with minerals, but neither had infinite resources. In particular, crude oil was a critical resource that was soon to show depletion problems. In 1970 US crude oil production reached its peak and started declining. That posed a critical strategic problem for the US government. Without an abundant supply of oil, the American empire risked the same decline that the British empire had seen just a few decades before, when it had passed its coal peak. The solution to the problem was found in the control of the still abundant resources of the Middle East.

The United States had relied on Middle East resources for a long time. In 1945 President Roosevelt met with King Ibn Saud of Saudi Arabia and seeded an alliance that lasts to this day. As discussed by Michael Klare in his book *Blood and Oil*,[48] this strategic vision continued with the oil crisis of the 1970s and was stated most clearly in the so-called Carter Doctrine expressed in President Carter's 1980 State of the Union address (and perhaps actually written by Zbigniew Brzezinski, national security advisor at that time[49]):

> Let our position be absolutely clear: An attempt by any outside force to gain control of the Persian Gulf region will be regarded as an assault on the vital interests of the United States of America, and such an assault will be repelled by any means necessary, including military force.

This statement is eerily similar to an earlier one on coal made by the British government in 1903.[50] In the 1970s and 1980s, the Soviet Union tried in various ways to match the US foothold in the Middle East, but without success. In 1988 Soviet oil production started to decline, and without the ability to control external sources of oil, the Soviet Union collapsed soon afterward. These events should not be seen as simple cause and effect. Rather, a series of entwined factors related to peak oil led the Soviet social, economic, and political structures to collapse together with oil production.[51]

Much US foreign policy after the fall of the Soviet Union can be seen as a continuation of the Carter Doctrine. The first Gulf War (1991), the invasion of Iraq (2003), and other events in the Middle East have clearly been a manifestation of the need for the United States to keep a tight grip on the region and control its petroleum resources.

Today, Middle East oil resources still play a fundamental role in the world's power game. However, although abundant, not even these resources can be infinite. There is much debate on just how long the Saudi resources can last. What is certain, in any case, is that the Saudi internal consumption of oil is constantly rising, and that is gradually eroding the capability of the kingdom to export its oil abroad. Similar considerations hold for the other major Middle East producers. Iraq has recovered from the destructions of the 2003 invasion and is now emerging as a major player in the world's oil market. But Iraq's resources have been damaged by war, and the growing Iraqi economy is absorbing more and more of the national production. On the other side of the gulf, Iran seems to be having serious difficulties maintaining its earlier levels of production, in part due to the political difficulties it is facing. Those difficulties became apparent in the late 1970s, during the turmoil of the Iranian revolution and the fall of the shah, and may have been related to the impossibility of the country's oil production continuing to grow, as it had up to then. Other minor producers in the region face the same problems and difficulties. The Middle East has been producing oil for nearly a century now. We can't expect it to keep going at the same rate for much longer. But whatever happens, it is unlikely that the major military power of the 21st century, the United States, will soon lose its grip on the region, which is still fundamental in the world's power game.

Another factor starting to play a role in the world's strategic struggle is the gradual reduction of oil as the dominant energy source. While oil production has been approximately static during the past decade, coal production has been rapidly growing. If the present trends continue, coal will soon surpass oil as the main energy source in the world. King Coal is coming back.[52] This trend is again changing the strategic game: the Middle East is producing very little coal (less than 0.1 percent of the world production),[53] whereas the main producers are, in order of decreasing importance, China, the United States, India, Australia, South Africa, and Russia.[54] In a sense the return to coal sets back the strategic clock by a century. Although oil remains a key resource, it may be gradually losing importance in military terms.

But the return of coal is not the only strategic change under way, and it may be that soon all fossil fuels will become obsolete. The latest-generation weapons are largely based on light and nimble robotic systems. In the future these light weapons may pack a tremendous amount of destructive power, especially if it becomes possible to develop so-called fourth-generation nuclear weapons.[55] But the present trend is to use these robots as precision weapons that share little with the earlier weapons systems' brute-force approach of carpet bombing and wholesale extermination. Robotic weapons can be directed toward highly specific targets, destroying the enemy's command and control system.[56] In battle, robots don't need to carry around the weight of the armor of traditional systems; it may cost less to replace a robot with another one than to provide it with expensive protection. As a consequence the new weapons need much less fuel and could be engineered to run on electric power, which can be generated by sources other than fossil fuels, such as nuclear and renewable energy. Renewable energy plants are especially interesting in military terms, since they can be dispersed over territories in such a way as to offer a poor target for the enemy. The strategic vulnerability of renewable energy is even lower if the energy source is associated with the weapon itself, for instance in the case of a drone powered by onboard solar cells.

An even more drastic change of strategic perspective could be the result of the recent emphasis on cyberweapons, designed to take control of virtual space. In recent years the US government aimed the virus called Stuxnet against Iranian nuclear enrichment facilities.[57] It is too early to assess the effectiveness of such weapons, but if it is possible to take over the enemy's command and control system, then the war can be won without the need to fire a single shot. Cyberweapons need a very small amount of energy compared to conventional weapons; in fact, their energy needs are supplied by the enemy.

With the 21st century, the cycle of mineral resources in their military role may have turned completely around. The importance of crude oil is gradually being deemphasized, while the central strategic role may now revert to metals—resources badly needed for all the electronics that power robots and cyberweapons. Metals such as copper, gold, cobalt, tantalum, zirconium, indium, and rare earths, as well as minerals for semiconductors, such as gallium, have become key strategic resources. That shift completely changes the game of world domination in ways that, at present, are difficult to predict.

Though the strategic emphasis may shift to different mineral resources in the near future, it is clear that economies, in peace as in war, need both energy and mineral resources and that the competition for what is left to be extracted can only become more and more stiff. If we go back to the times of the Roman Empire, we see that the Romans didn't take the depletion of their gold mines with philosophical resignation. They tried as hard as they could to keep them producing, and the result was *ruina montium* ("ruin of the mountains"), as described by Pliny the Elder in his *Historia Naturalis*. The mountains of the Spanish region of Asturias still show the destruction wreaked on them by Roman engineers.

But what the Romans could do to their mountains with picks and hydraulic fracturing is very little in comparison with what we can do to our mountains with explosives and diesel-powered machinery. We are already destroying one mountain after another in order to get at the coal seams they contain. It is a process that is not soon going to stop, as the world's economy gears up to recover the last accessible ores on the planet. It is truly a war waged against the planet, a take-no-prisoners war.

It also is a war that cannot be won. In the long run the planet will recover from the assault of human miners, and the only possible casualties will be us.

PART TWO

THE TROUBLE WE'VE SEEN

The Bingham Canyon copper mine in Utah, so vast it can be seen from space, is the world's deepest open-pit mine. As demand for minerals intensifies, techniques to access them grow more and more aggressive. The easily accessed ores are exploited first, then mining operations move on to lower quality ores, which require much more energy to extract.

4

The Universal Mining Machine:
Minerals and Energy

Imagine that you are an astronaut stranded on a remote planet, a mere chunk of rock orbiting a faraway star. Your ship was badly damaged by the impact of landing, but fortunately your antimatter power plant is still working. So you have plenty of energy, but the problem is that you need a new ship. You can have your robots build one for you, but only if the right materials are available: metals, semiconductors, glass, ceramic, and more. You don't have the time or the resources to prospect for mineral ores on the planet, even assuming that there are any. But ordinary rock contains all the elements of the periodic table—just locked inside in extremely tiny amounts. So you have your robots build a universal mining machine that extracts ordinary rock from the planet's crust. It crushes it, heats it, and then transforms it into an atomic plasma. The ions in the plasma are accelerated by an electric field and then separated according to mass by a magnetic field. At the output, you have all that you need: each element neatly packed in its box. With time you can gather what you need to build your new ship and go back home.

That's science fiction, of course. But there is nothing that defies the laws of physics in the idea of obtaining mineral resources from the undifferentiated crust of a planet. If it is physically possible, then why don't we build a universal mining machine here, on Earth? We could use it to produce all the minerals we needed from ordinary rock, and we wouldn't have to worry about such things as supply security, prices, and depletion anymore.

Some economists seem to be thinking exactly in these terms when they say that mineral resources will never be exhausted.[1] They seem to believe that a universal mining machine could be actually built. Unfortunately, the idea is attractive in theory, but not feasible in practice. The limits to mineral extraction are not limits of quantity; they are limits of *energy*. Extracting minerals takes energy, and the more dispersed the minerals are, the more energy is needed. Today, humankind doesn't produce sufficient amounts of energy to mine sources other than conventional ores, and probably never will.

Energy and Mineral Extraction

The Earth's crust is said to contain 88 elements in measurable concentrations that spread over at least seven orders of magnitude. Some elements are defined as common, with concentrations over 0.1 percent in weight. Of these, five are technologically important in metallic form: iron, aluminum, magnesium, silicon, and titanium. All the other metals exist in lower average concentrations, sometimes much lower. Most metals of technological importance are defined as rare. The average crustal abundance of elements such as copper, zinc, lead, and others is below 0.01 percent in weight (100 parts per million). Some very rare elements, such as gold, platinum, and rhodium, exist in the crust as a few parts per billion or even less. However, most rare elements form specific chemical compounds that can be found at relatively high concentrations, called deposits, in certain regions. As we know, some of those deposits that are concentrated enough that we can actually extract minerals from them are called ores.

Mining ores is a multistage process. The first is the extraction phase, in which materials are extracted from the ground. Then follows the beneficiation stage, when the useful minerals are separated from the waste (also known as gangue). Further processing stages normally follow; for instance, the production of metals requires a smelting stage and a refining one. All these stages require energy. Table 4.1 lists the specific energy needed for the production of some common metals, together with the total energy requirement for the present world production.

From this table, we can see that the world's production of steel alone requires 24 exajoules, equivalent to about 5 percent of the world's total primary energy production (about 450 exajoules).[2] Also note that, today, we extract copper from ores that contain it in concentrations of 0.5 to 1 percent. The total energy involved is 50 megajoules per kilogram.[3] Using this value, we find that we need about 0.7 exajoules for the world's copper production. This is about 0.2 percent of the world's total energy production. Taken together, the data of the table indicate that the total energy used by the mining and metal-producing industry might be close to 10 percent of the total world energy production—an estimate consistent with other projections.[4]

Table 4.1 is a snapshot of a situation that keeps changing as we continue extracting minerals. In its early history, mining required only minimal amounts of energy, as it was mainly provided for free by geochemical processes of the remote past. For instance, finding gold in a river required only a pan as equipment, and the product—gold nuggets—came already pure and ready to be used. But as gold mining went on, we gradually ran out of these easy resources, and today we mine gold from deposits that contain just 0.01 percent of it, and that's very expensive. It is a general trend: as we run out of high-grade ores, we have to move to lower-grade ores. The trend is evident for all metals, as shown, for instance, for copper in figure 4.1.

Table 4.1. Energy Required for Production of Some Common Metals

Metal	Specific production energy (MJ/kg)	World production (Mton/year)	Total energy required (EJ/year)
Steel	22	1,100	24
Aluminum	211	33	6.9
Copper	48	15	0.72
Zinc	42	10	0.42
Nickel	160	1.4	0.22
Lead	26	3	0.08

Note: EJ = exajoules (1 quintillion joules); MJ = megajoules (1 million joules); Mton = million metric tons.

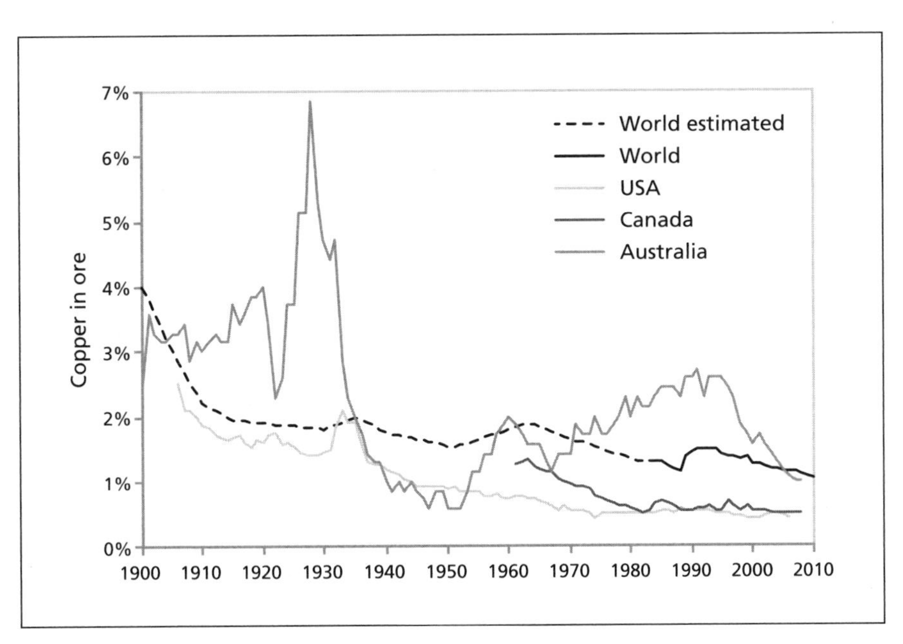

FIGURE 4.1. Dwindling grades of copper ore being extracted. *Note:* Rise in ore grade in Australia from 1972 onward is due to startup of the high-grade Olympic Dam mine.

As we move along this path, the amount of resources that we can theoretically access depends on Lasky's law, which says that the grade of an ore is inversely proportional to its abundance in the crust. In other words, low-grade deposits of a certain mineral are much more abundant than high-grade deposits. As a result, we see the counterintuitive result that the amount of extractable resources increases as extraction progresses because the industry is forced to extract from lower grade deposits. Curiously, the cake seems to become bigger as you eat it. That makes some people very optimistic about the future prospects of mining. A statement about crude oil attributed to Professor Peter Odell of the Erasmus University of Rotterdam, in the Netherlands, summarizes this attitude well: "We are not running out of oil, we are running into it!"[5]

Unfortunately, no matter how impressively large the amounts of dispersed minerals existing inside the Earth's crust, the problem lies in the large amounts of energy needed for extraction. In general, the lower the ore grade, the more energy is needed for extraction. For example, if an ore has a mineral concentration that is 10 times lower than another, it will take 10 times more energy to extract that mineral from the ore.[6] This is an approximation, especially when applied to the whole production process that includes smelting and refining. But we can take it as a reasonable "first order" approximation.

We saw that we are already committing about 10 percent of the world's primary energy to the production of minerals. This amount can only increase as we access lower-grade resources, even if we are aiming at just maintaining the present production levels. Therefore, if we want to maintain the current fraction of energy allocated to the mining industry, we must increase the world's total energy production in proportion. That has been possible, so far, by increasing the production of fossil fuels, but it is becoming more and more difficult. The problem of dwindling ore grades occurs also with fossil fuels; energy is becoming more and more energy-expensive to produce. Nevertheless, the extra energy needed to access low-grade ores must come from somewhere, and at present it is being drawn from other sectors of the economy. That can't be painless, and the pain appears in the present trend of rising prices for all mineral commodities.

For energy-producing resources, the problem of dwindling grade can be described in terms of energy returned on energy invested (EROEI).[17] EROEI is the ratio of the energy that a particular resource will produce during its useful life to the energy invested to access that resource (find it, build a plant, maintain it, recycle or dispose of it, and so on). Obviously, the higher the value of the EROEI, the better an energy source is. Energy costs and gains do not translate directly into monetary costs and gains, but in general there is a proportionality between

Platinum Group Metals: The Vulnerable Keys to Emissions-Control Systems

Ugo Bardi and Stefano Caporali

Precious metals are often considered useful only for their decorative properties, or as currency. However, some precious metals have important technological applications, and their gradual depletion may have important industrial consequences.

Such is the case with the six "platinum group" metals: ruthenium, rhodium, palladium, osmium, iridium, and platinum. They are precious in the sense that they share some of the properties of gold and silver: that is, they are rare, expensive, and also chemically stable (which is why they are commonly referred to as "noble" metals). But unlike gold and silver, which are coveted for jewelry and currency, the main value of the platinum group metals lies in their unique chemical properties. They are of fundamental importance in chemistry, biology, and medicine as catalysts—that is, as substances that can stimulate chemical reactions that would not occur in their absence. Three of these metals—platinum, rhodium, and palladium—find their main application as catalysts for the automotive exhaust converters designed to reduce the harmful emissions of internal combustion engines used mainly by road vehicles. However, it is possible that progressive depletion could make these converters too expensive in years to come, and that could create a significant pollution problem.

Practically all road vehicles today are powered by the familiar internal combustion engine. Most of these engines use hydrocarbons as fuel (diesel or gasoline). When these fuels are burned inside the engine, they generate mainly water and carbon dioxide, two gases not normally considered pollutants. However, the combustion of hydrocarbons also creates small amounts of highly harmful substances, including unburned hydrocarbons, carbon monoxide (CO), nitrogen oxides (NO_x), and particulate matter, typically in the form of very small carbon particles (nanoparticles). Additives to fuels may create other dangerous materials in the exhaust. For instance, until not long ago, tetraethyl lead was a common additive to gasoline, creating a considerable lead poisoning problem all over the world. Fortunately, today lead additives are forbidden by law in most (although not all) countries of the world.[8]

Starting with the second half of the 20th century, various filters were developed to reduce the emissions of harmful substances from the engines of road vehicles. For diesel engines, the filter focuses mainly on removing particulate matter, and it does not normally use precious metals as catalysts. For gasoline-powered engines, instead, the filter focuses mainly on eliminating CO, NO_x, and unburned hydrocarbons. This is accomplished by three-way catalysts based on platinum, rhodium, and palladium. Rhodium catalyzes the elimination of nitrogen oxides by reduction, while palladium catalyzes the elimination of carbon monoxide by oxidation. Platinum catalyzes both. This technology turns out to be very efficient and has become a fundamental factor in abating pollution from road traffic in urban areas. When in good condition and operated properly, the three-way catalytic filter can remove up to about 90 percent of the three noxious gases.[9]

On average, an automotive catalytic converter can store 1 to 3 grams of platinum and smaller amounts of rhodium and palladium. As a consequence, automotive converters now use more than half of the world's mineral production of platinum.[10] That raises the question of whether there exist sufficient mineral resources of platinum group metals to satisfy the demand for the foreseeable future.

The platinum group metals are all very rare in the Earth's crust. Production is concentrated in a few mines in South Africa, Russia, Canada, the United States, Poland, Zimbabwe, and Australia. Of these, South Africa accounts for about 85 percent of the total world production and has 82 percent of the world's resources.[11]

According to the United States Geological Survey, the total reserves of platinum group metals amount to some 66 million tons.[12] The current reserves-to-production (R/P) ratio points to a supply of about 130 years. This result would seem reassuring, but the R/P ratio is a poor indicator of the availability of a mineral commodity. The question is not for how many years we can theoretically produce these metals but how and if it will be possible to keep production at the present levels at reasonable costs. Because of the gradual depletion of high-grade ores and the increasing costs of the energy needed for extraction and processing, platinum prices increased fivefold from 1992 to 2012, reaching an all-time high of about $1,500 per ounce—more than $50 per gram. Additionally, the growth trend in world production stopped in 2005 and has been in decline ever since. That may cause prices to rise even more in the future. In order to reduce the problems brought on by high cost and declining availability of these platinum metals, we can consider the following strategies:

- Reduce the amount of catalyst in automotive converters
- Develop catalysts that are not precious metals
- Recycle platinum metals more efficiently
- Use engines that don't need precious-metal catalysts at the exhaust point

Reducing the amount of catalyst in the converter is possible by making the catalytic particles smaller, but there are limits to this approach. Below certain dimensions, the particles either lose catalytic capacity or are carried away from their substrates by the exhaust. It is also possible to vary the ratio of the different metals in the catalyst, for instance by partly replacing platinum with the less expensive palladium—that particular mechanism is being explored but doesn't, of course, solve the problem at its roots.

Developing nonprecious materials that can catalyze the three reactions of interest turns out to be a difficult task. Since the mid-1980s alternatives have been intensely investigated,[13] but a viable solution has not been found. Oxides such as perovskites[14] and boehmites[15] have been proposed as replacements for platinum group metals, but they still seem far from industrial applications.

Recycling can also counter depletion. Recovering precious metals from automotive converters is technically possible and economically convenient, especially in view of the current high prices of these metals. In fact, high prices have generated a brisk black market for stolen catalytic converters that find their ways to recycling facilities, proving the old adage that things done illegally are done most efficiently. Nevertheless, there are limits to recycling. Some cars are discarded too far away from recycling facilities and, in any case, the recycling process itself cannot be 100% efficient.

A further limit to recycling efficiency comes from the fact that precious metals are gradually lost during a vehicle's operation. One study estimates that a car's converter loses 6 percent of its precious metals after 80,000 kilometers.[16] In practice, the end-of-life recycling rate of platinum from catalytic converters reaches a global average of only 50 to 60 percent,[17] which is clearly not enough to "close the cycle" and solve the depletion problem.

So, to address depletion, we need to consider completely different approaches, such as using engines that don't require precious-metal catalysts at the exhaust point. One such approach would be to use fuels not based on hydrocarbons. Pure hydrogen (H_2) and compounds of hydrogen and nitrogen (such as ammonia, or NH_3) can power an engine, and the resulting exhaust would not contain unburned hydrocarbons, particulates, or carbon monoxide. The remaining problem of nitrogen oxides could possibly be solved without precious-metal catalysts. Such engines, however, have not found practical uses up to now.

Or we could eliminate internal combustion engines altogether. Electric motors are lighter, more durable, and more efficient, and they emit no pollutants during operation. The problem is, of course, how to obtain the electric power that these motors need. Some vehicles (e.g., trolley buses) can be powered by aerial wires, but for most road vehicles electricity must be generated on board. A possible way to do so is by means of fuel cells, devices that can use the chemical energy of fuels—typically hydrogen—for the direct generation of electric power without the need to use a thermal engine and a generator. Unfortunately, the kind of fuel cells normally considered suitable for road vehicles (that is, polymer electrolyte fuel cells) need about 1 to 3 grams of platinum catalyst per kilowatt of engine power. That translates to about 100 grams per car with the currently accepted power range.[18] So powering the world's present car fleet with fuel cells would be simply impossible given the constraints on platinum production and reserves, at least with the current fuel cell technology.

A better way to power electric road vehicles may lie in a new generation of automotive batteries that use lithium, a metal that is relatively abundant in the Earth's crust and may have considerably fewer depletion problems than platinum group metals. A move in this direction would not only greatly reduce pollution but also lengthen the life span of the presently available mineral resources of platinum group metals. So it turns out that, in this case, the depletion of a fundamental resource, namely the platinum group metals, is a problem but also an opportunity to move toward a better and less polluting technology.

the two. We'll delve deeper into EROEI in the next chapter, but it should be clear that it is a fundamental parameter in determining the ultimate limits of what we can extract and produce. For non-renewable energy sources, the value of the EROEI becomes smaller with the ongoing exploitation of the higher-grade resources, and in the long run it must become smaller than one, when the energy source ceases to be such and becomes a sink of energy. We are not yet there with our fossil fuels, but clearly it is a destiny we will face sometime in the future.

If fossil fuels offer little hope for a return to the past energy wealth, perhaps other sources could come to the rescue. Maybe a new generation of nuclear technologies or a rapid growth of renewable energy might invert the negative tendency. Would all depletion problems then be solved? In the short run, probably yes, but eventually we would face a fundamental problem: Lasky's law is just a rough approximation. Considering the complex processes that have created mineral deposits, it seems at least unlikely that a proportionality as simple as Lasky's law would hold. Geologist Brian Skinner has proposed that the distribution of minerals in the crust is bimodal, meaning that there is a large peak for the element at low concentrations in ordinary rock and a much smaller peak for the same element in deposits. The absence of concentrations between the two peaks is what Skinner terms the "mineralogical barrier."[19]

There are, of course, also exceptions to this rule. Uranium, for instance, does not seem to have a double concentration peak in its deposits, although this point is contested.[20] Then, of course, common minerals such as iron exist in high concentrations all over the crust and don't have a real mineralogical barrier. But even for iron, we don't mine the undifferentiated crust; we still mine ores. We could be facing some kind of mineralogical barrier if we were forced by depletion to switch from the currently used ores to different ones.

So, even if we could have relatively abundant energy for mining, eventually we would reach a point where there was little or nothing for us to mine. It is clear that, at some point, the only way to reach new sources of minerals will be to cross the barrier and mine the "other side," the undifferentiated crust. If we could mine in that region, we would have immense resources available. The problem is that the amount of energy needed is enormous, to say nothing of the tremendous environmental damage that would be done.

Take the case of copper, for instance. Copper is present at very small concentrations, about 25 parts per million, in the upper crust. To produce 1 kilogram of copper from the undifferentiated crust, we would need to process 40 tons of rock. We would need to break down rock at the atomic level, using about as much energy to destroy the rock as it took to form it. On average, that translates to roughly 10 megajoules (MJ) per kilogram, and so we can estimate that it would take about 400 gigajoules (GJ) per kilogram to extract copper from the crust, with the very optimistic assumption of a 100 percent efficient process. That's a lot of energy. The average American home consumes about 9,000 kilowatt-hours per year of electric energy, or 32,400 MJ. In other words, the cost of the energy needed to produce just 1 kilogram of copper from the undifferentiated crust could pay the average home electric bill for more than 10 years! Now, consider that we produce about 15 million tons of copper per year and you can understand what the problem is.

Compare this result with the energy needed to extract 1 kilogram of copper from the presently exploited ores, which totals about 50 MJ, and you have another way to understand how big the problem is: extracting from the undifferentiated crust requires an energy increase by a factor of ten thousand in comparison to the present needs.

You could say that looking at the energy needs of just one element is misleading, since a universal mining machine would produce all the elements together for the same energy expenditure. Still, if copper is representative of the increase in energy needed, and if we can't allocate more than 10 percent of our primary energy to mineral production, we still have to increase total energy production by a factor of about one thousand—far removed from anything we can imagine in the foreseeable future. In addition, the waste created by this kind of mining would run into the trillions of tons of rock per year, and damage to the ecosystem would be mind-boggling. The prospects of a universal mining machine are not bright.

Clearly we won't make much progress if we think we can solve the problem of mineral depletion by the brute-force approach of mining from ever decreasing ore grades. Could we think of a more subtle approach? Could we find more ores, different kind of ores, or completely different resources from which we could obtain the minerals we need? This is a question that deserves to be discussed in detail.

The first point to consider is whether we really know the amount of conventional ores in the Earth's crust. Here, of course, there are large uncertainties in the estimates, but it is unlikely that we could find substantial new resources. The Earth's surface has been thoroughly explored by mineral prospectors. Antarctica is the only major continent still unexplored for mineral resources, and there are most likely ores there. But at present finding or extracting anything that exists under kilometers of ice is an unthinkable endeavor. Maybe global warming will clear the ice away, but that is likely to take at the very least several hundreds of years, and it would bring a host of problems more serious than mineral depletion, including a sea level rise of at least 60 meters.

Could we just dig deeper for more ores? Not a good idea. First, it is terribly expensive. Then, ores form as the result of a variety of geochemical processes, most of which are active at or near the surface, and that's where we have been mining up to now. Maybe some special minerals could be found at great depths, but it is not likely that this approach could solve the depletion problem.

There is, then, the possibility of replacing conventional mineral sources with "nonconventional ores." There have been numerous ideas and proposals in this sense. Here too, however, we see that the basic problem remains the same: nonconventional ores require a lot of energy to be extracted and processed.

Mining the Oceans

The oceans contain large amounts of minerals, both in deposits in the sea floor and as ions dissolved in water. This fact inevitably leads to the question of whether it is possible to mine the sea floor.

The "sea floor," as a whole, takes several different forms. The bottom of shallow inner seas and lakes is normally similar to the surface of the continent in which they are located. In the case of the oceans, the sea floor begins with the continental shelf, which, geologically, is part of the continent it is attached to. At some distance from shore, the continental shelf drops down toward the deep sea floor (also referred to as the "abyssal plain"). The slope that connects the continental shelf and the deep sea floor is called the continental margin.

In terms of mining, the bottom of shallow seas and the continental shelf may contain mineral ores similar to those found inland. These ores may have formed underwater, as could be the case with crude oil. Or they may have formed during periods when the sea floor was actually above water, as may have happened for various regions of the continental shelf during the ice ages of the past million years or so. There is no doubt that mineral resources exist in these areas, but accessing them is not easy. Although the continental shelf is never at a depth of more than a few hundred meters, underwater mining requires complex and expensive technologies. The high costs involved may be justified only in the case of very valuable minerals, such as offshore diamond mines. That is done, for instance, off the coast of Namibia.[21] In some cases it is possible to mine undersea deposits as an extension of conventional mines, as is done in Japan for some coal mines.[22] It is often possible to extract oil and gas from the continental shelf because the process of offshore drilling can be completely automated and is not much different than it is on land—except for the need for a floating platform for hosting the drilling equipment. Of course, this kind of drilling carries risks that are not seen on land, as when the Deepwater Horizon drilling platform operating in the Gulf of Mexico exploded in 2010, releasing huge amounts of oil into the ocean ecosystem.

However, the Deepwater Horizon rig operated at a much greater depth than is typical of rigs located on the continental shelf, and that factor contributed to the difficulties the operators had in stopping the spill. The rig was looking for oil in the continental margin, a geologically active area that forms as sediments from the continental shelf cascade down its slope and accumulate in an area called the continental rise. This area is especially interesting for oil and gas prospecting but requires deep or ultra-deep offshore drilling, meaning drilling at depths of 3,000 meters and more. With the progressive depletion of conventional oil, deep and ultra-deep sources are becoming more and more important, but their amount is limited and the cost of extraction is very high, to say nothing of the risks of major spills involved.

A completely different case is that of the deep sea floor, also called the abyssal plain. The geology of this region is not the same as that of the continental crust. The ocean floor is formed by the geological "conveyor belt" that transports material from the oceanic ridges to subduction zones. It is continuously renewed and thus relatively young in geological terms—no more than about two hundred million years old and often much younger than that. (The continental crust, in comparison, may be billions of years old.) Most of the deep sea floor is geologically quiet and doesn't show the hot geochemical processes that create mineral ores on continents. Oil and gas could, theoretically, form on the deep sea floor, but normally the sedimentation rate of organic matter is low, and besides, oceans are sufficiently oxygenated that the dead organic matter is removed by bacterial activity before it can be buried. So most of the deep sea floor contains no oil and no gas.

But not all the deep sea floor is so quiet. The situation is very different at the mid-ocean ridges, where hot magma is continuously transported from the mantle to the surface. This rising magma carries to the surface metal ions dissolved in hot seawater percolating underground. When this hot water cools down at the surface, it releases those ions, typically in the form of sulfides. The process forms chimneylike vent structures, composed mainly of iron sulfide compounds. These chimneys may contain gold, copper, silver, and other metals.[23] Ancient chimneys that were once part of the seafloor—like the copper ores on the island of Cyprus—have been mined on land. But mining these deposits at the bottom of the sea, at depths of thousands of meters and far from any land, would be extremely expensive. Besides, these deposits are normally of a lower grade than most land-based hydrothermal deposits because the latter have often gone through secondary concentration processes that can take place only on land (with some exceptions[24]).

Nevertheless, some of these minerals accumulate. Relatively common in some areas of the deep sea floor are manganese nodules, which also contain iron and copper. There were some attempts to exploit these nodules in the 1970s, but with time the interest died out.[25] In general, sea floor deposits are too dispersed and at concentrations too low to be commercially interesting, even without considering the energy and monetary cost of mining at such great depths.

There is also another completely different possibility for mining the oceans: that of directly extracting the minerals dissolved in water as ions. In the 1920s, German chemist Fritz Haber looked at the possibility of extracting gold from seawater, but his attempts were a failure. Gold does exist dissolved in seawater, but in amounts so minute that extraction is practically impossible in macroscopic quantities. That doesn't mean that it is impossible to extract minerals from seawater, and indeed, it has long been done with some high-concentration ions, such as sodium chloride, or common table salt. But most metal ions in seawater exist in very low concentrations and have never been extracted in commercial quantities. However, the idea of extracting rare metals from seawater became popular in the 1970s, when a number of studies were performed on the subject in view of the rising prices for all mineral commodities. The idea was abandoned with the decline in mineral prices, but today it has returned. Then as now, though, no low-concentration metal is being commercially extracted from seawater.

The problems with extracting minerals from seawater are twofold: the limited amounts available and the energy requirement. Calculations of these parameters are not encouraging.[26] The oceans are vast, but rare metals are dissolved in them in extremely tiny amounts. In the case of copper, for instance, there is about 1 billion tons of it in the form of copper ions dissolved in the whole mass of seawater on the Earth.[27] That may seem to be a large amount, but consider that we now produce about 15 million tons of copper every year. Even if we were able to filter the whole mass of all the oceans—an unlikely prospect (also very bad from the viewpoint of fish, whales, and all other sea creatures)—we would run out of oceanic copper in little more than 60 years.

Of course some ions are found at higher concentrations and would have less extreme extraction requirements. However, even for the best case—that of lithium—in order to maintain the present production we would have to increase by a factor of 15 the amount of seawater being industrially filtered today in desalination plants. Again, this unreasonably assumes a 100 percent efficient process.[28]

These numbers give us some idea of the size of the task and of the tremendous impact seawater extraction would have on marine ecosystems. But those would be minor problems in comparison with the real one: energy.

Extracting ions dissolved in water doesn't require the energy-expensive process of rock breaking, lifting, and crushing of conventional mining. However, the concentrations of rare metal ions in seawater are enormously smaller than they are in mineral ores. So extracting a specific ion from seawater requires filtering enormously large amounts of water. That is not just a practical problem; it takes energy to pump water through a filtering membrane or, alternatively, for all the operations needed to transport the membrane to sea, leaving sea currents to move water in and out, and then to recover it. The second strategy may require less energy than the first, but in both cases we are talking of huge amounts. Even in the most favorable case—again, that of lithium—it is possible to calculate that even for a 100 percent efficient membrane it would take about 10 percent of the present world production of electric power to keep lithium production at the present level using seawater extraction.[29] For all the other metals dissolved in seawater, the energy requirement would be far, far larger.

The energy problem is especially critical if we consider the extraction of uranium from seawater—something proposed in the 1960s—as a solution for the uranium shortage that would have resulted from the great expansion of nuclear plants planned at that time.[30] Today, the stasis of the nuclear industry has made this problem less important, but uranium extraction from seawater is still discussed as a future possibility. However, it is possible to calculate that the energy needed to extract and process uranium from seawater would be about the same as the energy that could be obtained by the same uranium using the current nuclear technology.[31] That, of course, would make extraction from seawater useless. Perhaps, if more efficient nuclear technologies could be developed, then uranium from seawater could be a possible energy source, since we would need smaller amounts of uranium. But at present there is no practical interest in uranium extraction from seawater.

In short, with only the possible exceptions of lithium and uranium, extracting minerals from seawater in amounts comparable to the present production from ores is impossible. (See "Lithium: The Next Car Fuel?") That doesn't mean that oceanic water could not be a useful source of minerals if we were to limit our needs to smaller amounts. In this sense, some experiments with algae show promise.[32] If we were to be able, in the future, to use more efficient industrial processes, then it would be possible to use the oceans as a recycling system for those resources that cannot be completely recycled on land.

The Philosopher's Stone

Some ideas for new sources of minerals appear remote in terms of practical applicability but are still worth a glance. Could we think of creating the elements we need using nuclear reactions? This idea is equivalent to that of the "philosopher's stone," the dream of ancient alchemists: a way to transform lead into gold. It is not impossible to transform one element into another; in fact, it is done all the time inside nuclear power plants and particle accelerators. Heavier elements can be created from lighter ones by neutron capture, while lighter elements can result from the successive decay of activated nuclei. Nuclear fission—that is, the breakdown of atomic nuclei—can also generate lighter elements from heavier ones.

The equipment needed for these nuclear reactions is very expensive, but in a nuclear power plant those costs are paid for by the energy production. It is for this reason that plutonium is an economically viable fuel: in a certain sense, it comes for free as a by-product of energy production from uranium fission. If it were possible to generate plutonium in large amounts, it could even replace uranium; this was the idea fueling hopes for a plutonium-based economy in years past. However, those hopes were largely abandoned, in part because the special breeder reactors turned out to be prohibitively expensive and complicated and in part because of the risks involved in handling and managing plutonium. Today, all the world's reactors produce just about 70 tons of plutonium per year, by far too little to support a whole economy.[33] In comparison, about 380 tons of the fissile isotope of uranium (U-235) is produced per year from mines, and this amount is insufficient to fuel even the present fleet of nuclear reactors. But could we think, at least in principle, of using nuclear

Lithium: The Next Car Fuel?

Emilia Suomalainen

Until recent times lithium was known mainly as a dietary supplement to regulate human mood. But the appearance of a new generation of lithium-based batteries has changed everything. With electric cars appearing as a nonpolluting alternative to the oil-based dinosaurs we insist on using, lithium has become a crucial commodity for the transition to a cleaner world. A basic question remains, though: Do we have enough?

A soft and silvery-white metal, lithium is the lightest of all metals and the least dense solid element under standard conditions. It is an excellent conductor of heat and electricity and a highly reactive element present in traces in all organisms. It is not rare in the Earth's upper continental crust, but while there are a large number of lithium deposits, very few of them are of any commercial value as they are largely either too small or of too low a concentration.[34]

Lithium is used in a large variety of applications, from ceramics and glass to lubricating greases, desiccants, continuous casting, air purification, primary aluminum production, polymers, and pharmaceuticals. Its use as a primary energy source in fusion reactors has been discussed since the 1970s, though we are still far off from this application. But if lithium is not yet an energy source, it surely is an important energy carrier as the main component of both disposable lithium batteries and rechargeable lithium-ion (Li-ion) batteries.[35] Lithium use in secondary or rechargeable batteries has increased significantly in recent years as these batteries have become more and more popular in portable electronics and automotive applications.

Today, hybrid and electric vehicles (EVs) are still mostly powered by lead batteries or by nickel-metal hybrid (Ni-MH) batteries, but the use of the lighter, less bulky, and more efficient Li-ion batteries is rapidly rising despite their higher costs. With the diffusion of these batteries in large numbers, costs are expected to go down.[36] Other lithium-based batteries, such as lithium-sulfur and lithium-air, are expected to provide even better performance in the future.

The future demand for lithium in electric vehicles depends on several factors, notably global population growth, the development of passenger car markets in developing countries, the future people-per-car ratio, and the market penetration rates of electric vehicles. A large-scale transition to electric mobility would increase lithium demand dramatically. To meet this demand, one scenario projects that extraction rates would have to rise from today's 200 ktons per year to over 1,400 ktons per year by around 2050.[37] This would be a huge increase, comparable to the explosion of crude oil production in its early days, and it is not clear whether mineral production of lithium would be able to match such a large demand.

Sources and Production

There are three main types of lithium sources: brines, minerals, and seawater. Brines are saline waters that contain a great amount of dissolved salts. They can usually be found at locations where water (either freshwater or seawater) has undergone extreme evaporation, although geothermal and oil-well brines also exist. Brines formed by evaporation are commonly found in salt flats, the largest of which are situated in South America (Chile, Bolivia, and Argentina) and in China and Tibet. The highest concentration of lithium in brine resources can be found at Salar de Atacama in Chile.[38] This salt flat is also the world's largest currently exploited lithium deposit, producing almost 40 percent of the world's lithium.[39]

Brines are pumped to shallow solar evaporation ponds where secondary elements and compounds such as magnesium and sulfate are eliminated under controlled conditions. Lithium is finally recovered in the form of lithium carbonate. The use of "free" solar energy in the evaporation process greatly reduces the energy requirements for production and is the main reason brines are today's major lithium source. While Salar de Atacama currently produces the greatest amount of lithium, the world's biggest deposit is situated at Salar de Uyuni in central Bolivia. This deposit is currently unexploited because of its high altitude and limited potential for solar evaporation, along with other technical difficulties (the deposit has a high magnesium concentration), various environmental issues, and opposition from local communities. (The site, if exploited, would compete with local farms for a limited freshwater supply.)

The second major lithium source is solid ores such as pegmatites—igneous rocks containing lithium-rich minerals. Currently, lithium extraction from pegmatites remains expensive compared to using brines. However, in addition to lithium, pegmatites can contain recoverable amounts of other scarce elements such as beryllium, tantalum, tin, and niobium. Lithium has also been found in hectorite clays occurring at several locations in the western United States and in jadarite mineral deposits discovered in the Jadar River valley in Serbia.

The US Geological Survey (USGS) currently estimates that global lithium reserves amount to 13 million tons.[40] The USGS's reserve and the resource estimates have been increasing significantly in recent years: reserves (deposits with a high likelihood of extractability) more than tripled from 2009 to 2011, while the identified lithium resources (deposits with varying degrees of probable extractability) rose from 14 million tons to 33 million tons over the same period.[41] Chinese reserves have increased more than sixfold, Australian reserves have more than tripled, and Chilean reserves have more than doubled. These increases should be taken as a sign of uncertainty as to how large the reserves and resources are and whether their exploitation is currently or potentially feasible.

Some studies indicate that the USGS estimates may be conservative, while others propose much smaller estimates.[42] In any case, the amount of in situ resources is usually much higher than the actually recoverable amount of lithium.

Seawater also contains a substantial amount of lithium: the lithium concentration in seawater is about 0.17 parts per million (ppm), giving a total resource of about 2,500 billion tons, which is several orders of magnitude larger than the amount contained in land-based mineral reserves.[43] Lithium is one of the few minerals abundant enough in seawater that extraction is considered to be a concrete possibility, and indeed, both Japan and South Korea have made plans for such an exploitation. However, the process is currently uneconomic due to its energy requirements as well as other technological challenges.

According to USGS statistics, the major lithium-producing countries are currently Chile, Australia, China, and Argentina. These four countries produce almost 95 percent of the global output. These same countries also control the greatest lithium reserves, amounting to more than 98 percent of the world total.[44] Lithium production is therefore much more geographically concentrated than, for instance, the production of crude oil, where the four major producers provide only 40 percent of the global output.[45] The 2012 world lithium production was reported by the USGS to be 37,000 tons.[46] This value excludes the production of the United States (data are withheld to avoid disclosing company proprietary data), but the US contribution should not have amounted to more than 5 tons. In any case, the recent production trend is of a sustained increase.[47]

Do We Have Enough?

At the current production rate (37,000 tons per year), known lithium reserves (13 million tons) would last for more than 300 years. If, in addition, we could exploit all the estimated land-based resources, then we would have about a millennium's supply. On the basis of these numbers, all fears of lithium depletion would appear misplaced at present. Furthermore, if a practical and inexpensive technology for lithium extraction from seawater can be developed, we could say that lithium is "forever," not only because the amount available in the oceans is very large, but also because the sea would act as a reservoir, collecting the lithium we dispersed in the environment.

However, today we are limited to ground reserves, and they may not be sufficient for large-scale electric mobility (e-mobility, via battery-powered EVs) if the use of lithium in automotive batteries generates a considerable increase in demand. Exponential growth can significantly shorten the production lifetime: at a yearly growth rate of 3 percent, the global reserves would last a little more than a century, and at a 10 percent growth rate, the estimated lifetime would be less than 50 years.

These considerations have generated a lively debate on the adequacy of the lithium supply for large-scale e-mobility.[48] At present, the positive assessment of lithium availability seems to be gaining the upper hand. However, mobilizing the lithium necessary for e-mobility would require major investments in geological and engineering efforts,[49] and at present, these investments do not seem to be on the agenda. In addition to economic aspects, increased lithium extraction would undoubtedly have energy and environmental costs, notably in the form of the destruction of the unique Salar de Uyuni environment in Bolivia.

Note also that most e-mobility scenarios generally assume a high level of lithium recycling and recovery (80 to 100 percent). At the moment, lithium recycling is almost nonexistent: according to a recent study, the global recycling rate stands at less than 1 percent.[50] This dismal performance stems in part from the technical difficulties of recycling lithium from car batteries, but also from the fact that lithium recycling is not currently economically feasible compared to inexpensive primary lithium. However, recycling could become increasingly important in the long term, and some experts contend that it will be a necessary option—perhaps not for the buildup of lithium stock in EV batteries, but for the maintenance of this stock.[51]

On the whole, despite the reasonably good prospects of e-mobility, trading an energy dependence on oil for a material dependence on lithium does not seem a wise path for future transportation systems, especially because of the extensive and large-scale changes in infrastructure required by a shift from internal combustion engine vehicles to EVs. In addition, this shift would involve a strong dependence on the small set of lithium-producing countries. All in all, e-mobility needs to be seen in a larger context of green mobility along with other sustainable transportation options such as cycling, walking, car sharing, and greater use of public transport. The demands for mobility are diverse and the solutions are likely to be so as well. However, as some kind of road mobility is likely to remain important in the future, lithium batteries are also likely to play an important role as the "fuel" of nonpolluting vehicles. Hence, the management of lithium production and lithium recycling are important challenges we face for the future.

reactions to create rare elements in amounts sufficient to replace dwindling mineral resources? This possibility has been discussed since the early years of the nuclear industry.

One possibility is to exploit spent nuclear fuel, which contains small amounts of precious and valuable metals that, theoretically, could be recovered.[52] However, mining spent nuclear fuel is an extremely difficult, dangerous, and expensive process because of the radioactivity involved. The spent fuel can be reprocessed to generate new nuclear fuel, but it has never been possible to process it to extract minerals of commercial value. Even if it could be done, the total mass produced would not exceed the mass of the fissioned isotopes, which today is less than 500 tons per year of the fissile isotope of uranium (a little more comes from plutonium fission). These are extremely tiny amounts in comparison with those typical of the mining industry.

The outlook for neutron capture is better. We are already using the process to create materials that have commercial value. Technetium and americium are examples of unstable elements that don't exist on Earth but are created in nuclear reactors for their special properties. Technetium is used as a radioactive tracer in medicine, while americium is an ionization source in smoke detectors.

But could we do more than that? For instance, could we make the dreams of ancient alchemists come true and create gold by nuclear reactions? It is possible. One way is to irradiate an isotope of mercury (Hg-196) with neutrons.[53] The result is the unstable Hg-197 isotope, which decays into gold. It has been done, but unfortunately, in terms of replacing rare mineral resources, it is not such a great idea, as mercury can hardly be defined as a common mineral. But the main problem is that the amounts that could be created in this way are tiny, at best. Since each fission generates about 2.5 neutrons, the total produced today in the world corresponds to just around 5 million moles of neutrons (a mole, or gram-molecule, is a unit used in chemistry; each mole contains a very large number of atoms or particles, usually written as $6x10^{23}$). If we could exploit all these neutrons and if the reaction were to proceed at 100 percent efficiency, then we could produce a maximum of about 1,000 tons of gold per year. At present, the total world production from mines is around 2,000 tons per year, so in comparison, the amount theoretically producible by nuclear reactions is not so small. However, it is unthinkable that we could utilize more than a few percent of the produced neutrons, so at the very best we could create something like 100 tons of gold, and probably much less. Similar yields would result in any attempt to create platinum by irradiating iridium with neutrons.[54] So the amount of any element that we could create using the nuclear reactors we have today would hardly exceed a few tens of tons. Even if we were to greatly increase the number of nuclear reactors in the world, at most we could produce a few hundred tons per year of any element. These are amounts so small as to be negligible.

Though the prospects for using nuclear fission to produce mineral resources are poor, it is not impossible that the future will see the development of new and more powerful neutron sources not based on nuclear fission.

Today we already have a variety of such devices, including the so-called dense plasma focus, which initiates nuclear fusion.[55] The present-day technology cannot be used to create large amounts of materials by neutron capture, but after all, all the elements existing today on the Earth's crust were created long ago by neutron capture in supernova explosions. So who knows? One day the alchemist's dream could become true, not just as an exotic physics experiment but as a practical way to create useful materials. But we can't count on it to solve our present problems.

Mining the Solar System

The idea of mining other planets, asteroids, and other extraterrestrial objects is a pervasive theme of science fiction and often raised as a potential source of minerals, but it turns out to be little more than a dream. Even assuming that astronomical objects contain mineral deposits, the energy cost needed to reach them, mine ores, and then bring back the mined materials to earth is truly out of this world.[56]

Nevertheless, the high energy cost of mining astronomical objects might be overcome by moving there or perhaps bringing asteroids close to Earth by some kind of advanced propulsion device. Of course, these ideas involve gigantic technical problems, but in principle nothing that would be physically impossible. In terms of mining, however, there remains the fundamental problem that most bodies of the solar system just don't contain useful minerals. Earth's ores come from processes generated by a living planet, but most of the astronomical objects we can consider as mining targets are dead, both geologically and biologically.

The planetary body closest to us is our own moon. It is geologically inactive; it never showed plate tectonics and it appears that it never hosted liquid water on its surface. The composition of lunar rock was found to be not so different from that of the average earth crust, and therefore whatever could be obtained from the moon can be obtained here at a much lower cost. There exist proposals for mining the moon for a certain special mineral: the isotope of helium known as He-3, which collects there, brought by the solar wind, and could be used as fuel for nuclear fusion plants. But we are talking about a fuel that presents enormous recovery difficulties and can only be used for a technology that doesn't exist today and that we can't be sure will ever exist.

Asteroids have the same lack of ores, despite the fact that they are frequently mined in science fiction. Some asteroids could be good sources of

Nickel and Zinc: Twin Metals of the Industrial Age

Philippe Bihouix

Nickel and zinc are two metals emblematic of the industrial age. Both are used primarily to fight against the corrosion of iron and steel, but their physical characteristics make them useful for plenty of other purposes as well, such as for electricity storage in batteries. Not only are nickel and zinc used in similar ways, but they also face the same problem: exploitable deposits exist in limited amounts, and so the problem of depletion cannot be ignored. As is the case for copper, tin, and silver, the expected lifetime of nickel and zinc reserves is just a few decades, leading to a possible production peak quite soon. And it will not be easy to find sustainable and cheap substitutes.

Most of the nickel produced in the world (about 60 percent) is used in stainless steel. Stainless steel can have either a chromium base or a nickel-chromium base, but three-fifths of all stainless steel is the nickel-chromium variety.

The second largest application of nickel is in special alloys (that are up to 90 percent nickel) employed in harsh and high-temperature environments, as in aircraft engines or pipes in steam generators of nuclear reactors. Nickel can also be used in other kinds of alloys or as an anticorrosion coating material.

Finally, 10 percent of nickel production is consumed in various other applications, whether as a catalyst (to produce nylon or to hydrogenate oil for margarine), in various alloys for coins, or in chemical form in rechargeable batteries and as an additive in some glasses, paints, and plastics.

The largest demand for zinc, accounting for about 50 percent of production, is for galvanizing, or coating, steel to prevent it from rusting. Galvanized steel is employed in many industrial sectors, from construction and infrastructure to transportation.

Zinc is also used in zinc-based alloys, in brass and bronze (alloys with copper), and in a "pure" form (for instance as roofing plates). The zinc-based alloys and brass are resistant to corrosion, suitable for molding and die casting, and quite pleasing in color—three qualities that make them useful in almost all types of consumer products, from kitchen appliances and automobiles to household goods and devices.

Last but not least, zinc is used in the form of oxide for hundreds of applications—including as a pigment in paints, dyes, inks, textiles, and cosmetics and as an accelerating agent in rubber production.[57] It is also alloyed with manganese for use in disposable alkaline batteries, which represent more than three-quarters of the 40 billion batteries sold (and mostly thrown away) every year on Earth.

The Outlook for Reserves

Zinc and nickel have a similar abundance in the Earth's crust, ranging from about 70 to 80 parts per million. More than 12 million tons of zinc are mined each year, which makes it the sixth most used metal. Nickel is mined at a rate of about 1.8 million tons each year, putting it in tenth place. Considering their respective prices, their weight in the economy is comparable, with a global market of between $20 billion and $30 billion each.

Zinc ore is very often found together with lead ore, and it is also the main source of cadmium, germanium, and indium. Resources are spread around the world in about 350 mines, with ore grades (zinc content) typically between 4 and 20 percent. The most important zinc-producing regions are China (32 percent of the world's zinc production), Latin America (21 percent), Australia (12 percent), and North America (12 percent).[58]

Zinc demand is expected to grow at more than 5 percent per year according to industry forecasts, despite the current economic crisis. But the geological availability is limited: the average ore grade decreased from 7 to 5.5 percent between 2000 and 2012. Some major mines, like Brunswick in Canada and Century in Australia, will enter into decline soon, and new mines have higher operating costs because of the lower grade of the ores to be exploited. The "official" current reserves represent about 20 years of production, which should lead to tensions on the zinc market in the coming years and decades if nothing changes on the demand side.

The outlook for nickel reserves seems better than for zinc. The reserves-to-production (R/P) ratio yields estimates of about 45 years of supply at the current production rate. With new techniques (like hydrometallurgy) allowing nickel to be extracted from ores with nickel concentrations of 1 percent or less (as opposed to the 2.5 to 3 percent that was the norm a few years ago), we may be able to extend the production span to 80 or 100 years, but no more than that, and with the problem of increasing energy costs for extraction.

For both nickel and zinc, new resources could be found in the polymetallic nodules lying on some areas of the ocean floor. These nodules contain mainly manganese and iron, but they are also about 1 to 1.5 percent nickel and copper, while hydrothermal polymetallic sulfides are rich in zinc (typically 5 to 15 percent).[59] If it were possible to exploit these resources, our reserves estimates for extractable nickel could be approximately doubled and so would extend the lifetime of the resource a few decades, but at a much higher price and energy expense.

We can expect, then, that nickel and zinc production will be affected by a general "peak everything" effect, which will be caused by the increasing costs of extraction of fossil fuel resources, which are necessary for the extraction of almost everything else. The easily exploited part of the reserves has been already removed, and so it will be increasingly difficult and expensive to invest in and exploit nickel and zinc mines.

Can Recycling Make a Difference?

The problem with recycling most metals, including nickel and zinc, is that we do not use them only in a basic, metallic form. We also use them in dispersive or dissipative applications. When zinc oxide is used in toothpaste, for instance, it won't be recycled in the water treatment unit. Nor will it be reclaimed when it is used as a white pigment or an additive in plastics or glass. When used in car tires, infinitesimal quantities of zinc are left on the roads; the rest is disposed of in landfills or mixed in ashes when old tires are incinerated. More than 5 percent of the zinc we use is directly dissipated with such applications. Nickel has similar dissipative applications, such as when it is used to fix dyes in the textile industry or for yellow pigments, but its loss due to dispersal is smaller (probably around 1 to 2 percent).

Even when used in bulk metallic form, nickel and zinc are lost at prodigious rates in the waste stream, such as when they are discarded in landfills or end up in incinerators. It is possible to improve on separated collections for recycling, but the design of consumer goods does not help in this task: we handle every day a vast number of different materials and alloys—up to three thousand just for nickel. And we can't keep track of every paper clip or staple we use.

Finally, there is the effect of downcycling, which is especially apparent in the case of nickel-containing steel. Stainless steel and high-grade nickel alloys are generally collected and recycled, as nickel is an expensive metal considered worthy of collection. Stainless steel scrap is then turned back into stainless steel (sometimes by adding some primary nickel to increase its grade). On the contrary, low-grade alloys (containing small amounts of nickel) and steel parts that are only plated with nickel are not separated for collection and are considered low-value steel scrap. So they are mixed with other carbon steel, and the nickel becomes even more diluted. The output is low-performance recycled carbon steel; the nickel has been physically recycled, but from a functional point of view it has been lost. About 15 percent of nickel is "recycled" in this way.[60]

Similar downcycling effects occur with zinc, with the additional problem that zinc is volatile at high temperatures and large fractions of it disappear in furnaces during melting. Recent legislation has imposed methods to recover the metal lost in this way, but they are not applied worldwide and are not 100 percent efficient.

So, even though 55 percent of nickel and 35 to 40 percent of zinc is recycled, about 45 percent is lost at each cycle. In other words, if you start with 100 kilograms, only 30 kilograms will remain after two cycles. Depletion comes fast.

The Long View

Can we imagine a world without nickel or zinc?

Of course we will never completely run out of these metals. We have extracted 50 million tons of nickel since the end of the 19th century, and there are probably 35 to 40 million tons still around in our infrastructure and buildings. We still have an additional 80 to 100 million tons underground, a sizable portion of which we will extract and store in various objects in years to come. These metals will remain with us for a long time.

So the real question is, for how long can we support our current consumption? And what will we do if our consumption exceeds our capability of extraction? If a general collapse of economy and population occurs, the remaining population will have plenty of metals. Humans may enter a post-industrial age, not as hunter-gatherers but as metal scrap gatherers. But such a solution to scarcity would be drastic and surely unpleasant for those who have to go through the transition. Are there technical solutions to reduce the impact of scarcity for these two strategic metals?

Substitution for both nickel and zinc is theoretically possible in many cases. An abundant metal such as aluminum can be used as an anticorrosion coating, though it involves more expensive processes. Protective coatings could also be made using organic layers, plastics, or paints, but these have less mechanical resistance than metals. Titanium, also abundant in the Earth's crust, promises good natural resistance to corrosion and might be used in several applications where stainless steel is now used. However, titanium's high melting point makes its processing expensive. And wholly new technologies will be necessary if we intend to substitute for nickel in many specialty applications. For instance, there is no presently known substitute for the nickel "superalloys" used in high-temperature engines.

So a better strategy for the midterm would be to reduce the bleeding rate of nickel and zinc during their industrial cycle.

To reduce losses, we need to reduce or stop dispersive uses of these metals, and their use in disposable and short-lived objects. This process would include completely reviewing our waste management system and designing products that are less complex and easier to dismantle. To reduce the speed of the cycle, we must design and manufacture only repairable, reusable products; fight against obsolescence; and be cautious about our fascination for new things.

All these changes can slow the gradual loss of nickel and zinc; nevertheless, depletion is unavoidable, sooner or later.

One consequence may be that we will once again accept corrosion. We might decide to let simple agricultural or housing tools turn slowly into rust. For other objects, some regular painting or coating could be sufficient. Other applications will simply become impossible, particularly in the chemical, oil and gas, high-tech energy, and nuclear industries. (It is worth noting that the nuclear industry is one from which metals, being irradiated, cannot be recycled.) But when we get to a point of severe scarcity for nickel or zinc, abandoning such industries may not be such a big deal; we'll have much worse problems by then, and after having switched from cars to bicycles or horses, our thirst for special metals will be considerably reduced.

nickel, but nickel is just one of the mineral resources we need, and hardly the most important one. Comets could be considered good sources of water, but again, we have enough water in a place much closer to us: the Earth's oceans.

There are a few exceptions to the lack of ores on astronomical objects. The four largest moons of Jupiter, along with Titan, one of the moons around Saturn, are geologically active rocky bodies and may contain ores. Titan contains hydrocarbons at its surface, whereas the moons of Jupiter (with the exception of Io) contain water at or near the surface. Other bodies that have been geologically active in the past are Mars and Venus, and both might still contain ores formed during their active phase. Of all these cases, the only remotely conceivable target for human mining is Mars. If we ever establish a self-sufficient colony there, colonists might be able to exploit local ores for the mineral resources they need.

Overall, it is not impossible to conceive future scenarios in which breakthroughs in energy technologies allow humans to expand in the solar system and mine astronomical bodies. But it turns out that, likely, the planet best endowed with mineral resources is the one on which we stand right now: Earth.

Depletion Is Unavoidable

We have seen that most of the optimism regarding the depletion problem comes from a basic mistake: that of considering the *amounts* of minerals available and not the *energy cost* of recovering them. If we had low cost and nearly infinite energy, depletion would not be a problem; we could build a universal mining machine and recover useful minerals from anything at hand, whether ordinary rock or waste. But this is not the case, and only conventional ores can be profitably mined with the amounts of energy we can produce today. The abundance of other possible sources, from ions dissolved in oceans to the planets and asteroids of the solar system, is an illusion: these resources are, energy-wise, too expensive to mine. So in the future we'll have to face a progressively more important depletion problem, enhanced by the fact that our energy sources, mainly coal and hydrocarbons, are also subject to depletion.

Up until about 10 years ago, tins like this one containing black caviar from the Caspian Sea were common and inexpensive in Russia. Then they disappeared from the market as their source, the Caspian sturgeon, nearly disappeared from the sea. You can still find black caviar in Russia, but it is rare and hugely expensive. Caviar is an example of a fishery that has been exploited to near extinction. It is not the only case of a theoretically sustainable resource destroyed by overexploitation.

5

The Bell-Shaped Curve:
Modeling Depletion

At the beginning of the 19th century, the economic growth that came from the coal revolution generated a brisk demand for home lighting. The widespread use of kerosene was still decades away, and so most people met the fading light and evening hours with oil lamps, a technology that was thousands of years old. The only alternative was town gas, generated by the gasification of coal, but providing that was a complex and expensive process, possible only in large towns. Everywhere else, some kind of liquid fuel was needed to keep the oil lamps burning, and with populations growing and demand increasing, the traditional vegetable oil or animal fat sources had become even more expensive. So, when it was discovered that whale oil could burn with a clean flame and that it could be produced at low cost, the whaling industry boomed.

By the mid-19th century whaling had become a major worldwide industry that employed entire fleets and produced more than 10 million gallons of oil per year, most of which was used as lamp fuel.[1] By the time Herman Melville published *Moby Dick* in 1851, whale oil was to lamps what gasoline is to cars today: it powered almost all of them. But while Melville recounts life aboard a whaling ship and shows the ferocity with which whales were pursued, he fails to tell us explicitly that the whale-oil industry was already in decline by that time. The overexploitation of the whale fisheries had depleted the stocks, and the species being hunted at the time had become rare.[2] Some modern studies indicate that for one species, the right whale, only about 50 females remained in the oceans at the end of the 19th-century whaling cycle. The production of whale oil went through a bell-shaped curve. It peaked around 1845 and never recovered afterward. Prices increased, and less and less oil was available. Fortunately for lamp users, kerosene was by then being distilled from crude oil and quickly replaced the dwindling whale oil supplies.

It may be that the mad rush for whale oil portrayed in *Moby Dick* was a symptom of the difficulties the whaling industry had at the time. Whalers may not have wanted to admit it, but they were running out of whales.

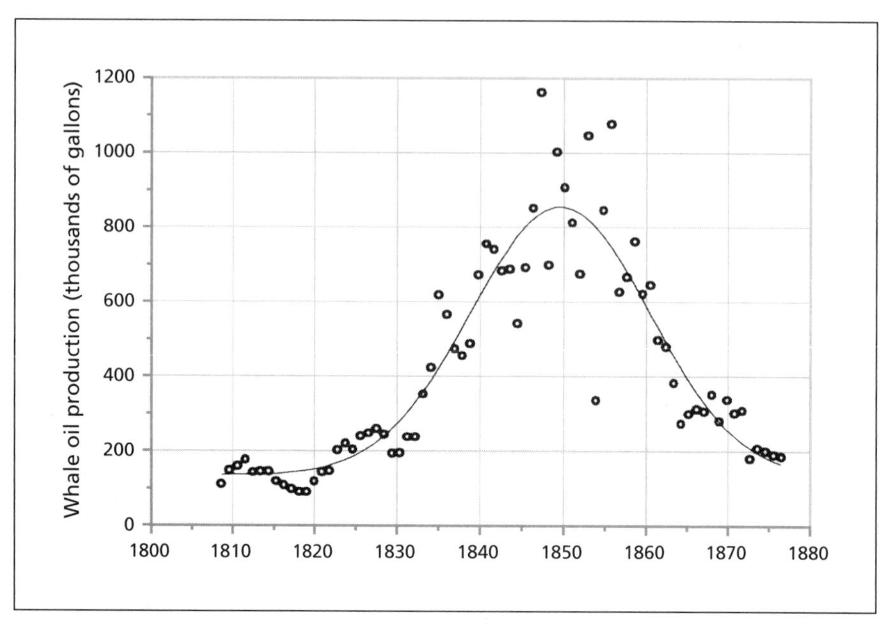

FIGURE 5.1. Whale oil production in the United States.

Whaling is not the only example we have of a fishery that ran out of its stock because of overexploitation. Modern examples abound. The caviar trade nearly wiped out sturgeon in the Caspian Sea, where catches peaked around 1980 and declined afterward, leaving the fish critically endangered.[3] The northwest Atlantic cod fishery collapsed in the 1990s. In fact, overfishing has put about 85 percent of the commercial fishery species at serious risk. All these cases show that human exploitation is perfectly capable of destroying even theoretically renewable resources.

The Bell-Shaped Curve

The story of the 19th-century whaling industry offers us an example of a nearly complete cycle of exploitation of a natural resource that started at zero production and ended at zero production, when the resource had been nearly completely consumed and exploitation made no more sense from an economic point of view. Whales are, obviously, renewable in the sense that they can reproduce. But they can do that only slowly, and in practice they were destroyed much faster than they were able to reconstitute their numbers. So the bell-shaped curve of the whaling cycle has many of the characteristics of the exploitation of a nonrenewable resource, such as oil or coal. In this sense the historical data of the production and price of whale oil provide for us a precious "laboratory" of how an industry based on a nonrenewable resource operates and how the cycle develops until there exist no more resources to exploit.

Examining historical cycles of resource exploitation, we can find a large number of cases where the production curve is bell-shaped and symmetric, like it was for whale oil. We can find at least one example that is even older than the 19th-century whaling cycle: the rise and fall of timber production that led to deforestation in Ireland. Once again, a renewable resource was consumed at speeds much faster than it could renew itself. In Ireland, as everywhere in the world, trees were an economic resource much sought after. John Barrington, an 18th-century Anglo-Irish landlord, once remarked, "Trees are stumps provided by nature for the repayment of debt."[4] The ancient Irish forests were destroyed by the late 18th century, when less than 1 percent of the island's surface maintained trees.[5] Deforestation in Ireland had especially tragic consequences. Trees take a long time to regrow in the cold Irish climate, and bare soil is easily subjected to erosion by rain. Ultimately, the loss of fertile soil was an important factor in generating the famines that started in 1848 and killed more than a million people.

Modeling Depletion

The first attempt to build a model that would describe resource depletion came with *The Coal Question* of 1856 by William Stanley Jevons.[6] Jevons examined coal production in light of a basic principle of economic theory: that of diminishing returns. The cost of coal extraction varies depending on such factors as the quality of the coal, its depth, and the thickness of the vein. The easy coal is, obviously, extracted first, and that makes coal progressively more expensive to produce. Jevons concluded that depletion would eventually make coal too expensive for the British industry to afford. At that point production would start declining. Jevons didn't propose a "bell-shaped" curve, as Hubbert would do about one century later. But his line of reasoning was certainly compatible with such a concept.

The Coal Question was an advanced study for its time but had only a modest impact on the later development of resource economics. Possibly because depletion was a problem for the far future, the problem lay dormant until it was approached again after World War I, when the availability of mineral commodities had become a crucial strategic problem. Harold Hotelling was probably the first economist to propose a quantitative model for the depletion of finite resources. His model, developed in 1931 and known today as Hotelling's rule,[7] was destined to have a strong impact on economic thought.

Hotelling's model is based on the concept that the owner of a mineral resource, assumed to be finite, has a choice about whether to extract it and sell it on the market or leave it in the ground. The owner might decide to extract everything immediately, sell the mineral, and invest in the stock market. Or the owner may decide that it is better to keep the resource underground and sell it later at a better price. The decision will depend on the perceived discount rate—in other words, on how the owner values a dollar gained in the future in comparison to a dollar gained immediately (a bird in the hand is worth two in the bush). Hotelling had to make some assumptions; an important one was that the owner had complete control over the resource and could decide at what price to offer it on the market. This condition was defined by Hotelling as that of a "perfect monopolist." At this point Hotelling demonstrated that the owner could maintain constant revenue from the mine if prices were increased exponentially while production was slowly decreased. Production would go to zero when the price of the resource reached that of its "backstop resource"—a once more expensive alternative that could replace the first resource.

An easier way to understand Hotelling's rule may be to think of beer cans in a refrigerator. Imagine that the cans cannot be replaced; in this case, each can will seem more valuable as fewer remain. As a consequence you'll tend to drink less as time goes on. Note, however, that the model is based on some very restrictive assumptions. For instance, using the beer example, you'll tend to drink less only if you are the only beer drinker in the house, a "perfect monopolist." If there are several drinkers, your best strategy instead is to drink as much as you can, as fast as you can, for as long as there is beer available. So we may imagine that Hotelling's model wouldn't work so well in the real world, and indeed, it doesn't. There are some cases where mineral resources have had exponential price increases, but in most cases prices have tended, so far, to decline or to show a U-shaped curve, where decline is followed by a sharp rise. As for the prediction that production should slowly decline, again, this has been the case for very few mineral resources; on the contrary, most mineral commodities have shown a continuous increase or, sometimes, bell-shaped production curves.

Hotelling's model was part of a general movement of ideas in the 1930s that sought to conserve natural resources. The model did show that depletion problems were to be expected in the future, but, rather optimistically, the presence of a "backstop" resource was always assumed to save the day. However, the model has been often misunderstood and forced to conclusions that it does not support. For instance, the fact that the prices of most mineral commodities have shown a declining trend up to recent times has been interpreted as implying that the resources were exploited only to a minimal fraction of the amount available.[8] Others concluded that the resources weren't limited at all. Exemplary in this sense is Julian Simon, who, in his book *The Ultimate Resource*, arrived at the conclusion that the worldwide mineral resources are "infinite" on the basis of five price trends.[9]

It goes without saying that Hotelling's rule cannot be used to support these views. The fact that the prices of nonrenewable commodities may go down with time is mainly related to factors that the model doesn't take into account, such as technological improvements and factors of scale. That doesn't mean that the model is useless; all models are approximate and all are useful as long as we know their limits. The general rise in prices of the world oil resources observed in the past decades could have been interpreted as a probable prelude to a decline if observed with "Hotelling's lens," but that was almost never done.

Hotelling's model predicts that mineral resources are inevitably destined to run out at some time. However, other economists working in the same period developed more optimistic models. Still today, the commonly held position in economics is based on a model developed in the 1930s that goes by the name of the "functional model" or the "resource pyramid."[10] This model starts from the same assumptions that Jevons had considered in *The Coal Question*: that extraction starts from the most profitable resources and then gradually moves to less profitable ones. According to Jevons, it is because of this phenomenon (and not because of the abstract reasoning of Hotelling's rule) that prices go up and production goes down. The functional model, instead, assumes that high prices will stimulate the development of new technologies that will lower costs. As a consequence, prices go down and production increases as low-grade resources, normally more abundant than high-grade ones, are exploited. In other words, in this resource pyramid we start with small amounts of high-grade ores (the tip of the pyramid) and move down toward larger amounts of lower-grade resources (the lower layers of the pyramid).

This functional model has some realistic elements, but it fails to account for a point that both Jevons and Hotelling had emphasized: over a certain limit, rising prices cause a reduction in demand, and that will eventually stop the rise in production. The industry just won't extract resources so expensive as to be impossible to sell. As a consequence, there is a limit to the kind of low-grade resources that the industry can exploit. The functional model sweeps this problem under the carpet by assuming that technology will always come to the rescue, lowering the costs of extraction and restoring both the demand and the profits of the industry. Unfortunately, this is a leap of faith: technology has monetary and energy costs, and there are limits to what it can do. And one thing is for sure: no technology can extract minerals that are not there.

Another model describing the relation of mineral resources and the economy is the one developed by Robert Solow in 1957.[11] It describes the production of economic goods as the result of a number of factors, including resources, capital, and land. All these factors are grouped together in a "production function," a mathematical expression that describes how each factor affects production. The production function can take various forms, and it can also include a parameter describing finite mineral resources, which are normally assumed to decrease exponentially with time as a result of depletion.[12] However, the effect of depletion is contrasted by a multiplicative factor, termed "Solow's residual," which grows exponentially and is supposed to describe the effect of technological progress. Adjusting the parameters, the model can be engineered in such a way that technological progress trumps depletion. The function describes the growth of the world's economy up to recent years. But projected to the future, it predicts that the output of the world's industrial system will keep growing forever, despite the dwindling production of mineral resources. Herman Daly, the economist who has spent decades disproving the myth of endless growth, summed it up best when he said that this approach is equivalent to saying that a cook can always prepare a larger cake with less and less flour available, simply by stirring the ingredients faster.[13]

Solow's model is often cited as a reason for optimism in assessing the future availability of mineral resources. It has been used as a major argument against the more pessimistic results of the models used for the *Limits to Growth* study of 1972.[14] There is no doubt that Solow's residual can generate never-ending growth on paper, but since the residual is not based on actual measurements, it finds little justification in physical reality and even violates the law of diminishing returns, a basic feature of most economic theories. Besides, it is possible to account for most of the increasing output of the world's economy by factoring in the increasing energy production devoted to extraction—something that refutes the need to resort to an arbitrary adjustable parameter.[15] Since energy is produced mainly from exhaustible resources, there is no reason to assume that growth will continue forever in the future.

The Tragedy of Mineral Commons

The main problem of conventional economic models in dealing with exhaustible resources is that they don't normally generate the often-observed bell-shaped production pattern. Hotelling's rule generates a continuously dwindling production, whereas Solow's model generates the opposite behavior, a forever growing production. If we want a description of the bell-shaped curve that reality has presented to us in many cases, we need to move to a different class of models, often generated outside the boundary of what is commonly recognized as economics. We can start examining these models with a well-known one proposed by Garrett Hardin (a biologist and not an economist) in his 1968 paper "The Tragedy of the Commons."[16]

Hardin describes a pasture that is the common property of a number of shepherds. That is, each shepherd can use the pasture without limits or extra costs. The question is how to optimize the exploitation of the resource (grass) in order to obtain the maximum amount of capital (sheep). There is a maximum number of sheep that can graze on a specific pasture. Exceeding this number means destroying the grass and creating a situation that, in modern terms, we define as "overexploitation." In the current, standard view of how free markets work, most economists would assume the optimal number of sheep would be reached by the work of the "invisible hand," the notion (proposed by Adam Smith in the 18th century) that each person acting in his own self-interest creates a set of conditions that improve the lot for everyone. In other words, improving your own bottom line is a win-win for all. However, according to Hardin, a problem arises when a number of independent operators, each one engaged in optimizing his or her gain, rely on the same resources.

Let's assume that every shepherd can decide how many sheep to take to the pasture. If we start with just a few sheep per shepherd, we may be well below the maximum yield that the pasture can provide. So each shepherd gains something by adding one extra sheep to his herd. In this way, at some moment, the total number of sheep grazing on this pasture will reach the maximum of sustainability. At this point the addition of extra sheep reduces the overall yield of the system. Unfortunately, from the viewpoint of each shepherd, adding one more sheep to his herd is convenient because the damage done will be spread over all the shepherds, while the gain will go to the single shepherd alone. Everyone reasons in these terms, and the overall result is that the number of sheep increases well above the maximum sustainable limit. Once that happens the pasture will be overgrazed and destroyed.

Hardin's model can be seen as the consequence of the failure of Hotelling's rule to take into account that in real life there is no such a thing as a perfect monopolist. The shepherds in Hardin's model behave like a group of beer drinkers who all get their beer from the same refrigerator. The best strategy for each drinker is not to save cans of beer for later but to drink as many as possible as fast as possible. The result is that the beer disappears fast and everyone is left without it.

Talking in terms of beers or shepherds is, of course, a highly simplified way to describe the real world. People are not always so ill behaved that they'll steal beers from each other, and there is no evidence that historical pastures managed as commons ever underwent the "tragedy" of overgrazing that Hardin described. Both at home and with pastures, there exist social brakes in the form of laws, habits, and peer pressure that prevent the rapid destruction of the resource being exploited, be it beer or grass. But if we examine the story of the 19th-century whaling industry described earlier on, we see that Hardin's model works beautifully. Whalers always reasoned in terms of maximizing their individual benefit; in other words, they acted according to the age-old principle of "grab what you can, when you can." No wonder whales were harpooned at the fastest possible rate.

The case of the whaling industry is not unique in the fishing industry, and the phenomenon of overexploitation of fisheries was discovered even before Hardin proposed his model.[17] In the jargon of economists, fish is a "free access" resource, and it cannot be optimized because no one can claim ownership of a specific fishery. The invisible hand fails to optimize the system, despite the fact that everyone operates to maximize his or her profits. One way to ease this problem would be to eliminate the very concept of "commons"—that is, to privatize the resource. In practice this is not always possible, especially for resources such as fisheries, as the sea can hardly be fenced. Privatization of natural resources also leaves the control of ecosystem services needed for the public good—like clean air, clean water, and ample fish stocks—in the hands of a private few, with potentially different motivations. One could argue that government intervention, regulations, quotas, treaties, and other measures to limit overexploitation can solve these concerns. But these measures have had limited success, and in modern times the overexploitation of fisheries has led to a worldwide, large-scale "tragedy of the commons."[18]

One might think that such overexploitation is linked to the difficulty that operators have in measuring the amount of resource available, but this doesn't seem to be the case. Estimates of stock sizes are usually available to fishermen, but that has not stopped overexploitation. There are many cases, in fact, where resources known to be in peril are nevertheless ravaged. Consider bison hunting on the American central plains in the 19th century. When large-scale hunting started, there were several tens of millions of bison in America. At that time, hunters could not have missed the fact that bison herds were fast disappearing, but they operated on the principle that if they themselves didn't kill as many bison as they could, someone else would. It was a perfect example of the tragedy of the commons at work. In a few decades, fewer than a thousand bison were left alive.

Can we apply the free-access model to mineral resources? In this case it would seem that Hotelling's rule should apply. Each firm exploiting, say, crude oil owns a certain number of oil fields and can decide how fast to exploit them. According to Hotelling, each company should gradually reduce production in order to maximize its revenues by exploiting the expected rise in prices. In practice this is not what we have been seeing. In exploiting oil fields, oil companies have been normally acting as whalers during the heyday of the whaling industry. That shouldn't be surprising; clearly oil companies are not monopolists in the oil market, as Hotelling's rule would assume them to be. Each company has a choice: it can optimize the economic yield of the fields it owns over a long time, or it can exploit the same fields as fast as possible in order to obtain a fast profit to invest in new fields. Only the second strategy can lead the company to growth, and it is the one normally chosen. For oil companies, the planet is a commons to exploit in pursuit of oil fields.

Hardin's "tragedy of the commons" tells us that the production of a resource should initially increase when exploitation is still in its early stages. Then, as the resource is consumed, its gradual destruction will invert the trend, generating a reduction in production. Between these two opposite phases should be a peak. Qualitatively, therefore, the model can be interpreted as producing a behavior similar to that of the Hubbert model. (See "The Hubbert Model.") But it is still just a qualitative model. How can we obtain something more quantitative?

Rabbits and Foxes: Modeling Overexploitation

In the 1920s Alfred Lotka and Vito Volterra independently proposed the first dynamic model of population evolution that turned out to have wide applications even outside biology.[19] It took into account only two types of organisms: predators and prey. In the simplest version of the model, the prey population tends to grow exponentially, as it is normal for biological species. However, the growing abundance of prey causes an increase in the population of predators. As predators take their toll, the number of prey stops growing and starts

The Hubbert Model: Looking Ahead by Looking Back

Marco Pagani and Stefano Caporali

There is much debate about just how large our supplies of mineral commodities actually are. The uncertainties are great, and the currently accepted modeling methods have led to estimates that are often way too large and sometimes too small. Is there a more reliable way to determine what we'll be able to extract in the future?

Forecasting the exploitation of natural resources is a task that has always been fraught with uncertainties and disappointment. Often the models used have been very simple, as when the productive lifetime of a mineral resource is estimated only on the basis of the reserves-to-production (R/P) ratio. In other words, the estimated reserves of a mineral are divided by its current rate of production to yield a projected timeline for extraction. It is an assessment that has little bearing on what the actual future will be like because there is no documented case in which the production of a mineral resource remained constant for a significant fraction of the exploitation cycle.

That's not the only problem with forecasting production. Perhaps a more important one is that the concept of "reserves" itself is fraught with uncertainty. "Reserves" defines what is considered extractable (or exploitable), but extractability is a property that depends on rapidly changing factors related to the economy. Yet it is often estimated by geologists on the basis of geological parameters. These two contradictory approach lead to great uncertainties.

There is, though, an approach that goes in a different direction—trying to estimate reserves not from geological data but from the historical production pattern. In other words, it projects the probable future by looking at the actual past. This approach has limits, too, but it can provide useful insights.

One key to the accuracy of the model is that it takes two basic realities into account: the first is that mineral stocks are nonrenewable; the second is that the most profitable resources are exploited first. To understand why these simple facts aren't always reflected in industry projections, we first must look at some industry definitions.

The terms *resource*, *reserve* and *ore* are often interchangeably applied to mineral deposits, but in fact they have distinct meanings.

An ore is a mineral or aggregate of minerals from which economically important minerals can be extracted. A concentration of naturally occurring ores whose economic extraction is currently or potentially feasible is called a mineral *resource*. The fraction of this resource that is economically and legally eligible to be extracted at a given time constitutes a *reserve*. Mineral reserves are extremely rare, and their discovery involves costly and determined efforts. Many reasons contribute to the failure of mineral deposits to be qualified as either ores or reserves, such as grade, size, depth, location, politics, and environmental concerns, among other issues.

Sometimes the development of new technologies allows the exploitation of low-grade or previously not extractable minerals, letting some resources become ores and reserves. For instance, worldwide copper resources were greatly increased at the beginning of the 20th century when new techniques allowed copper to be recovered from low-grade deposits. As another example, in the mid-1960s the generalized use of cyanide leaching allowed invisible, micron-sized particles of gold to be profitably extracted. Of course there are opposite examples, too. Ores may become too expensive to be mined; reserves can be downgraded to resources, as when newly issued legal and environmental constraints raise the extraction costs.

Such was the case at the Campiano mine in Italy, where mid-1980s prospecting showed evidence of a large pyrite deposit that contained interesting amounts of copper and zinc. Once the deposit was reached by means of underground works, however, it was realized that the copper and zinc content in the ore largely fluctuated, making processing very difficult. As a result the whole mining activity became uneconomic and the mine was abandoned.

The large degree of uncertainty in our knowledge of the underground can also pave the way for hoaxes, such as the one that began in 1993 when a small Canadian company, Bre-X, purchased a property called Busang in a remote area of Borneo. Bre-X began on-site exploration and reported increasing amounts of recoverable gold, from the initial 6 million ounces in May 1995 to 30 million in January 1996 and 71 million by February 1997. These increases attracted investors; the

Bre-X stock price rose from $0.08 to more than $210 per share. In March 1997 an independent analytical report stated that there was far less gold than expected and that there had been a "salting" of ore samples. The mining project was immediately abandoned, and the Bre-X executives fled to a foreign country to avoid prosecution.

Examples like these show why modeling past extraction rates can deliver reliable forecasting. This approach starts from the observation that many mineral resources show an exponential increase of their production rate over time. For instance, the production of metals such as copper, zinc, nickel, and platinum showed an exponential increase during the 20th century, with growth rates of 3 to 4 percent per year.[20]

But, of course, exponential growth cannot be sustained forever, a truth that wasn't recognized as a real problem until M. K. Hubbert showed how it applied to US oil reserves in the 1970s.[21] Hubbert pointed out that oil production followed a bell-shaped curve, ushering in an understanding of what was later termed "peak oil." From this observation, the procedure usually called Hubbert linearization was born. It is used to estimate the ultimate recoverable resource (URR) for crude oil. Plotting the annual reported production (p) as a fraction of the cumulative production (P) on the vertical axis and the cumulative production on the horizontal axis, the result is expected to be a straight line that intersects the horizontal axis at the value of the URR.

It can be expected, then, that the cumulative production of all mineral resources follows a similar pattern and that the Hubbert model can be used to more accurately project mineral reserves. Using the historical data provided by the US Geological Survey,[22] we have analyzed 10 transition and post-transition metals that illustrate a clear, single-peak behavior: chromium, molybdenum, tungsten, nickel, platinum-palladium, copper, zinc, cadmium, titanium, and tin. Other elements, such as lithium, antimony, cobalt, and iron, show two distinct peaks and cannot be treated with the simple analysis presented here, and others, like mercury, lead, and gold, show many oscillations linked to economic and political trends.

Figure 5.2 shows the Hubbert linearization plot for world chromium production. The URR estimation is 490 Mt, with a confidence interval between 360 and 660 Mt.

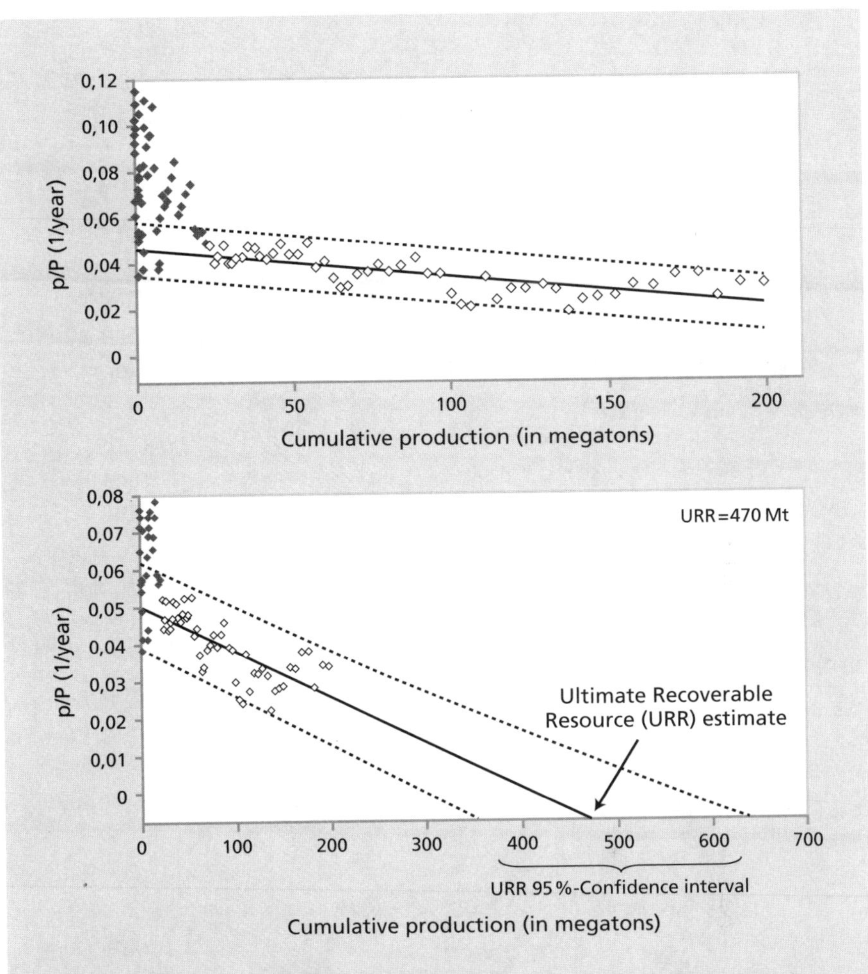

FIGURE 5.2. Hubbert linearization plot for world chromium production. This chromium analysis was performed on 50 data points, from 1962 to 2011 (open dots) plotting the ratio of the yearly chromium production to the cumulative chromium production (p/P) as a function of cumulative production (P). In the top graph, the solid line shows linear regression; the dotted lines show a 95 percent confidence interval. Gray points refer to the early stages of production and do not reflect the overall trend of the cycle. Hence, they were not used for the regression. In the bottom graph, the URR (ultimate recoverable resource) predicted value is determined by where the regression line (modeled on actual production trends) meets the horizontal axis (the cumulative production), together with its 95 percent confidence interval.

The USGS gives reserves data as single values with no indication of the uncertainty inherent in the determination (or in terms of confidence interval, the range of values with a given probability of being correct). However, it is possible to estimate the reliability of the USGS estimates by looking at how the URR values changed over time.

The case of chromium is quite sensational. In 2000 the USGS declared 3,600 million metric tons (Mtons) of world reserves, leading to a URR value of 3,730 Mtons, since 130 Mtons had been already extracted that year. In the following years reserve estimates were drastically decreased, going down in the range of 350 to 480 Mtons from 2010 to 2012, which corresponds to a URR between 530 and 680 Mtons. Comparing these USGS data to the result obtained by Hubbert linearization, we can say that the USGS is slowly moving toward a more realistic determination of chromium reserves.

The opposite case is observed for zinc, as shown in table 5.1, which compares the Hubbert and USGS projections for the URR of all 10 of the minerals we analyzed. The URR estimate in 2000 obtained from the USGS data was significantly lower (510 Mtons) than the Hubbert prediction (930 Mtons). In the following years this number increased slowly to reach in 2012 the lower bound of the confidence interval of the Hubbert estimation. The same happened for nickel and copper.

For other metals (cadmium, molybdenum, tin, titanium, and tungsten) the URR determinations given by the USGS over the last 10 years oscillate within the 95 percent confidence interval of the Hubbert estimate.

Table 5.1 shows that for cadmium, tungsten, molybdenum, chromium, and titanium the Hubbert prediction is lower (at 60 to 90 percent) than the value derived from USGS data, while for tin, nickel, zinc, and copper it is slightly higher (at 125 to 140 percent). In all cases the difference is not great, taking into account the year-to-year oscillations of USGS estimates. On average, the 95 percent confidence interval spreads from −20 percent of the mean value to +35 percent, with the exception of molybdenum, which has the worst value for the correlation coefficient. At the opposite end lies cadmium, which is the most depleted metal and has the narrowest correlation range in the prediction: from −12 percent to +12 percent.

There are also a few cases in which the Hubbert model fails to provide reliable results regarding URR. Rare earths fall in this category: the USGS values oscillate between 90 and 115 million metric tons (Mtons), while the Hubbert estimate is about one order of magnitude smaller.

Table 5.1. Ultimate Recoverable Resources (URR)

Metal	URR according to USGS		URR according to Hubbert			
	URR 2013 (Mtons)	Range 2000–2012 (Mtons)	URR (Mtons)	95% C.I. (Mtons)	Correlation Coefficient	Number of observations
Platinum	0.081	0.079–0.084	0.034	0.031–0.037	0.857	31 (1982–2012)
Cadmium	1.62	1.45–1.73	1.55	1.37–1.74	0.950	41 (1972–2012)
Tungsten	6.24	4.27–6.24	5.02	3.90–6.42	0.808	56 (1957–2012)
Molybdenum	17.6	9.64–17.6	15.60	9.90–30	0.543	38 (1957–2012)
Tin	25.0	23.4–26.5	30.80	25.8–36.7	0.861	60 (1953–2012)
Nickel	130	80.8–130	225	159–392	0.614	35 (1978–2012)
Chromium	665	533–3740	486	366–663	0.744	51 (1962–2012)
Zinc	707	509–707	935	742–1210	0.794	47 (1966–2012)
Titanium	1010	573–1020	868	663–1260	0.737	31 (1982–2012)
Copper	1270	736–1270	1900	1480–2600	0.734	44 (1969–2011)

Note: Shown here are URR figures determined with the Hubbert method and computed from USGS reserves data, for comparison.

There are several reasons accounting for this fault. First, rare earths have found large industrial applications only in recent years. Just a few years ago they were seen just as by-products of the extraction of more valuable minerals, and therefore their annual extraction rate was driven by the latter. Second, they do not have a decades-long established extraction industry such as copper and zinc have, and therefore their natural sources vary widely. They can be extracted from hard materials, such as phosphates (monazite and xenotime) associated with magmatic rocks that can extend deep into the Earth's crust, or from soft laterite clays or alluvial placers that are typical of surface environments. These differences contribute to make the evaluation of the exploitability of a rare earth's deposit more problematic. Third, recovery methods are still under development and the industrial processes are subject to continual renewal, leading to some fluctuations in the production rate. Finally, because they are elements of strategic importance to modern industry, true and affordable data about their production are not in the public domain.

But for most minerals the Hubbert model appears to yield more reliable results for the ultimate recoverable resources than other common models using USGS data. Too often the declared reserves are more hypothetical than measured; the data are affected by geological, political, or social nescience that has led, and in some cases is still leading, to the assumption of mineral bonanzas that are far removed from reality. Much more suitable results can be indirectly achieved by taking into account more verifiable and consistent sets of data such as the amount of the extracted resources or annual extraction rate.

dwindling toward zero. At this point it is the predators who are in trouble, starving and dying in large numbers because of the lack of prey. With predators nearly gone, the prey can start growing again and the cycle of boom and bust repeats. The model is based on two coupled differential equations and typical results can be seen in figure 5.3.

The Lotka-Volterra model does not pretend to describe the complex reality of actual biological systems.[23] However, it turns out to be a rich source of insight on the behavior of complex systems. In these systems, the interactions among the various elements occur by the mechanism called "feedback." Changes in a single element generate changes in all other elements—a characteristic that makes the behavior of these systems often unpredictable and always surprising. Indeed, it is well known that any outside intervention on a biological system usually generates a cascade of changes. Many attempts to rid ecosystems of pests, for instance, have resulted in unwanted changes, including the loss of species that were critical to that ecosystem.

These same feedback principles can be applied to human society as well. The economy, for instance, is a complex system of different entities (industries, customers, governments, and more) that interact with each other in a series of feedback relationships. These feedbacks sometimes generate growth or, at other times, decline or even collapse. In particular, the human economic system may be seen as a predator that grows on the availability of natural resources (the prey). In this kind of system, we can have oscillations like those seen in the Lotka-Volterra model. Just as the foxes cannot rationally limit their consumption of rabbits, humans cannot avoid prioritizing their short-term gains over their long-term ones. As a result, natural systems are exploited beyond their capability to regrow. When this happens (and it happens all the time) we reach overshoot, or overexploitation. The final result can be seen in the cycles of boom and bust so often observed in economies. When the resource being exploited is nonrenewable, there can only be a single cycle that leads to the total destruction of the resource.

Ultimately these cycles, both biological and economic, are based on energy: the amount of energy required to exploit a resource must be considered alongside the amount of energy gained from it. So again we return to EROEI, the energy returned on energy invested.[24] And, again, in the extractive industry we see that the tendency is to first extract the easy, high-EROEI resources that provide high returns. With time, the industry must move to progressively more difficult (lower EROEI) resources. As the returns diminish, less energy remains available for extraction, and the production growth slows. Eventually production peaks and then declines. Because minerals don't regenerate—except potentially in geological time—they don't experience the full feedback loop described by Lotka and Volterra in their biological model. In other words, the minerals, unlike prey, don't regenerate in the absence of predators. If these considerations are set in mathematical form, the result is the symmetric bell-shaped or "Hubbert" curve.[25]

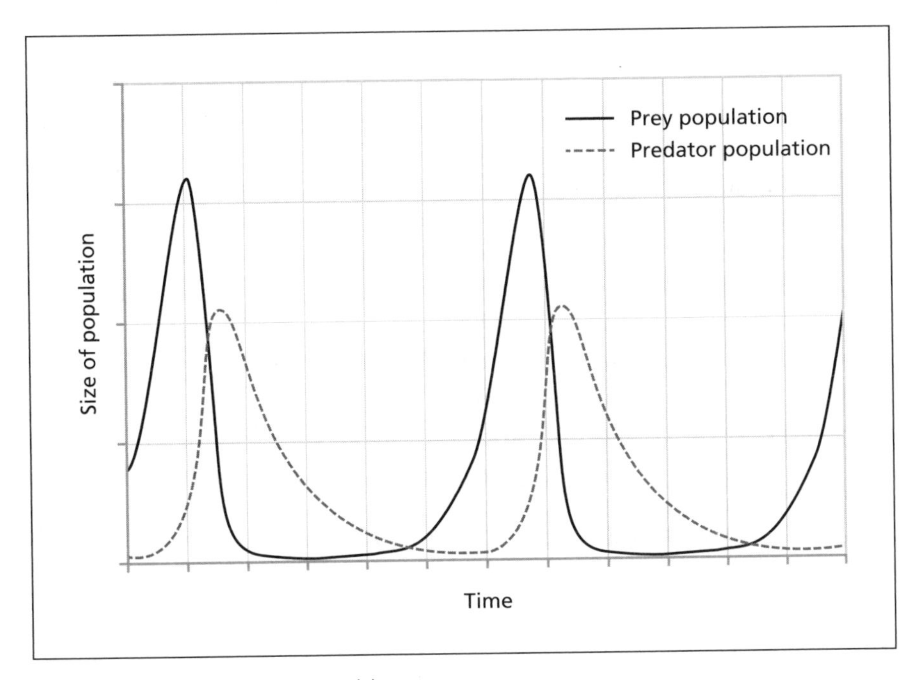

FIGURE 5.3. The Lotka-Volterra model.

Of course there isn't a rigid law that describes the behavior of the extractive industry. As mentioned earlier on, there are several cases in which the production of a mineral commodity has not generated a bell-shaped curve. Where we don't see such a curve, the model tells us that some factors other than free-market conditions are acting on the system. Because Saudi Arabia's ministry of petroleum regulates production with political and other opportunities in mind, we don't expect to see—and we don't see—Saudi Arabian oil production following a bell-shaped curve.

Production can also deviate from the expected curve under the effect of prices. The oil market is relatively inelastic, but not completely so, and the recent increases in oil prices have led to a rush to exploit every possible source of liquid fuel, mining unconventional sources like shale, tar sands, and the like. That trend has led output from some areas to grow once again after a decades-long decline. In some cases, as in Russian oil production, changes in political and economic factors can lead to a second cycle of exploitation, with a new phase of growth, peak, and decline. Nevertheless, the EROEI-based model generally describes the overarching tendency of the system. It tells us that, given some conditions, like the freedom for operators to choose where to invest their resources, resource production tends to follow a bell-shaped curve.

The Achilles' Heel of the Mineral Industry

The Lotka-Volterra model is the ancestor of an entire field of modeling, termed "system dynamics," that uses the same approach to describe more complex systems. We will examine later on how these methods can be quantitatively used to analyze the long-term possible consequences of mineral depletion on the world's economy. However, even at a qualitative level, the knowledge of the basic properties of complex systems allows us to examine some fundamental questions. For instance, we know that the world's economy depends on a large number of mineral commodities. Each nonrenewable commodity—be it oil, gas, uranium, or whale oil—is subject to an exploitation cycle that starts with zero production and must, eventually, end at zero production when the resource has been depleted to the point that an industry cannot afford to produce it any longer. This raises a big question: Will a critical resource become too insufficient for the industrial system that produces it and thus bring down the whole economic system?

The metaphor of Achilles' heel is often used when a large and apparently solid structure fails because of a critical defect. Petroleum could be the Achilles' heel of modern society. Indeed, how could we survive without the fuels manufactured from petroleum? Yet it may be more difficult than it would seem at first sight to bring down a complex society by the lack of a single critical commodity. Modern society is a complex system, not only because it is composed of many elements, but also because the elements are strongly linked to each other by a series of interacting loops that create a variety of feedbacks. A complex system will normally react to an external influence by rearranging its internal structure in order to minimize the effect of the external force.

Think of a city road system, for instance. If you close a road in a city, traffic will take different routes and continue to move, although it may slow down. You may close not just one road but several, and the transportation system may keep functioning, albeit at a reduced speed. The road system is "connected"; there are many alternative routes to the same place. Of course, after a certain critical point, the system will not be able to adapt anymore and will collapse in a giant traffic jam that freezes all transportation. Nevertheless, a city transportation system is enormously more resilient than, say, a mechanical watch. It is their very complexity that makes complex systems so resilient, despite their apparent fragility. Though the study of network resilience is in its early stages, network theory points to the same conclusions.[26]

So, if you look at mineral resources as a whole system, it is probably a mistake to think that some specific mineral will act as the Achilles' heel of the economy and bring it to collapse. That doesn't mean that depletion is not a problem; it means that the system can adapt to it—within limits, of course. The historical record seems to confirm this interpretation. The world oil crisis that started in 1973 is a good example of how resilient the system is. The US oil production peaked in 1971, and the rest of the world was unable to compensate. For some years global production actually declined and oil prices shot up. The disaster seemed to be irreversible at the time, but the system eventually adapted. High oil prices stimulated new investments in exploration and the development of new fields. New technologies were developed, and the use of oil was reduced or eliminated where it was not strictly indispensable. For instance, electricity generation was turned in large part to coal. After about 10 years, the crisis was basically over. Prices came down, although not to the same levels that had been commonplace before the crisis. Production growth restarted, although at a much lower rate than before.

We see a similar pattern whenever the production of a mineral resource pales in comparison with the need. The normal mechanism of price and production enters into play. A classic case is that of "solar grade" silicon. There simply wasn't enough available to support the rapid growth of solar cells starting around 2007. The dearth was due not to depletion (silicon is very abundant as a mineral) but to the lack of investments in plants able to produce the high-purity silicon necessary for solar cells. The market reacted with an increase in prices, which caused an increase in the cost of photovoltaic cells. But it was a temporary phenomenon. High prices stimulated new investments, and the production of solar-grade silicon increased to match demand, bringing prices down again from 2009 onward, to levels lower than they were before 2007.

A more recent case is that of rare earths, a mix of high-atomic-weight metals that are today a critical resource for a variety of applications, mainly in electronics. Rare earths are used in magnets, lasers, fiber-optic cables, mobile devices, and other electronics.[27] At present about 95 percent of the world's production comes from China.[28] (See "Electronic Waste and Rare Earths" on page 225.) It has been said that China has been thinking of its rare earth resources in strategic terms, just as some Arabic countries saw their oil resources during the oil crisis in 1973.[29] In September 2011 China announced the halt in production of three of its eight major rare earth mines, responsible for almost 40 percent of China's total rare earth production.

So will China strangle the world's industry with its monopoly on rare earth production? It is too early to say, but most likely the answer is no. Despite their name, rare earths are not so rare in the Earth's crust. An embargo from China would only cause the return to the exploitation of non-Chinese mines, and the system can adapt to the growing expense of rare earths by developing alternative technologies and methods. This point must have been clear to the Chinese authorities, since no embargo was enacted. The prices of rare earths rose to very high levels in 2011, probably as the result of the rumors about the embargo, but recently, they returned to values not much higher than they had been before the price spike.

It appears that in most cases, the world's industrial system is sufficiently resilient to adapt to a shortage of even very important resources such as crude oil. But that may not always be so; some resources might turn out to be really critical. One such case may be phosphorus, a mineral for which, in agriculture, there are no substitutes. The problem is described in "Phosphorus: Is a Paradigm Shift Required?" The lack of phosphorous, if not countered by radical changes in the way modern agriculture is managed, could be a true Achilles' heel of our society.

Phosphorus: Is a Paradigm Shift Required?

Patrick Dery

As a mineral, phosphorus does not have a mighty public profile. It does not seem as interesting, attractive, or crucial for the economy as, for instance, crude oil, gold, or rare earth metals. We don't hear of it wielding political or economic power. But phosphorus, in the form of phosphates, goes way beyond these other minerals in terms of importance for human life. We could live without oil and other fossil fuels—although we couldn't support as many people as we do now. We could live without gold—and probably we would all be better off for it. But we could not live without phosphorus: agriculture depends on it.

Farmers apply phosphate to enrich their soils, plants absorb it while growing, and in most cases those plants are shipped far away for consumption—carrying that absorbed phosphorus right along with them. In general, industrial agriculture makes greater use of mineral phosphates than the various forms of small-scale agriculture, including organic agriculture and permaculture, which try to recycle fertilizers. But the problem is always the same: once that phosphorus is displaced, the soil needs more. So without phosphates as fertilizers, agriculture—at least as most of the world currently knows it—could not exist. And without agriculture, what would we eat?

All this makes the general lack of concern, even among a number of agricultural experts, about a phosphorus depletion problem quite surprising. Let's consider some important and virtually uncontested facts about phosphorus. First, phosphorus is one of three essential macronutrients, or fertilizing elements, for plants, and also an essential element for animal nutrition. The other two are nitrogen and potassium. Phosphorus is second only to nitrogen as the most limiting element for plant growth. Crop yield on 40 percent of the world's arable land is limited by phosphorus availability.[30]

In agriculture phosphorus is essential for the production of symbiotic nitrogen from legumes (such as clover, vetches, alfalfa, and soybeans), which constitutes about 60 percent of all nitrogen used in agriculture. So when concentrations of phosphorus in the soil are low, it has a tremendous impact on symbiotic nitrogen production.

Given that there is no alternative to phosphorus, it can be considered life's bottleneck. Phosphorus can't be extracted from the atmosphere, as its common compounds exist only as solids. It is present in small amounts in ocean water but at very low concentrations, making extraction from this source too expensive to be practical. As a mineral commodity phosphorus is mined from the Earth's crust. Phosphate rock minerals are the only significant global resources of phosphorus.[31]

Most mined phosphorus must be converted to a soluble form before it can be used as plant fertilizer, much of which is lost after application. Only an estimated 20 percent of the phosphate fertilizer that is applied is absorbed by plants in the first year. That leads to large phosphorus loads on prime agricultural land and to troublesome runoff problems.[32] In developed countries the intensive use of soluble phosphorus fertilizers (like superphosphate, which is the result of acid treatment of phosphates) generates water pollution, saturates soils, and causes dangerous nutrient buildups in lakes and marine estuaries, where it can lead to oxygen-depleted (hypoxic) dead zones.

Moreover, phosphorus is often locked in the soil in chemical forms unsuitable for plant uptake. And most phosphorus is then exported (in products such as food, material, and fuel); it won't return to the land it came from because it is transferred to different geographic locations through a one-way economic system that converts mined resources to waste. These known and accepted facts are sufficient to clarify that there exists a major problem with phosphorus.

Phosphate rock production, like that of oil and other nonrenewable resources, follows a typical cycle. Production increases rapidly, often exponentially, at the beginning and then slows down before reaching a plateau or a peak in production. The peak is followed by a decline, and production may be completely stopped when it is no more economically interesting. It is undeniable that several regions of rock phosphate production have already reached their peak and are declining. The island of Nauru and the United States are good examples of this trend, as shown in figure 5.4.

The United States is the second largest producer of phosphate rock in the world. It has been surpassed by China (72 Mtons) and is

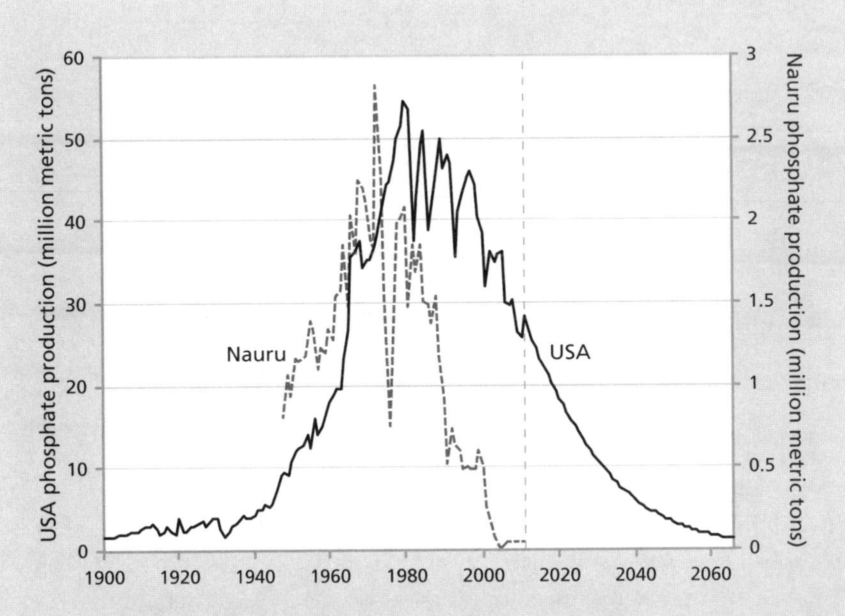

FIGURE 5.4. Annual production of rock phosphate. Phosphate rock production on Nauru has reached a near zero level (45 ktons extracted in 2011 from a peak of 2,823 ktons in 1973). Production in the United States has declined at a rate of 4 to 5 percent annually (28.4 Mtons extracted in 2011 from a peak of 54.4 Mtons in 1980).

being closely followed by Morocco and Western Sahara (27 Mtons). The top six global producers (China, United States, Morocco and Western Sahara, Russia, Jordan, and Brazil) represent approximately 80 percent of global rock phosphate production. Of these, four have been relatively stable for the past 12 years; one, the United States, is in steep decline. China shows the opposite trend. Having more than doubled its production since 2006, it currently represents 38 percent of world production. In the long run it is unlikely that China's reserves will be able to sustain this growth rate, but some countries, such as Morocco and Western Sahara, have a sufficient reserve base to be able to attain a substantial production expansion.

Worldwide, there remain large phosphorus deposits—more than 300 billion tons—but the vast majority of them are situated on the continental shelves and on seamounts (underwater mountains) in the Atlantic and Pacific Oceans. These resources are difficult to access and expensive to extract.[33]

The actual world reserve base is not well known, given the closure of the US Bureau of Mines in 1996. Hence, it is difficult to forecast the trajectory of rock phosphate production in the future. We can say, however, that the cheap and easily accessible phosphates are mined first and that, in the future, new production will come from nonconventional sources—those that have lesser concentrations of phosphorus. These deposits will require more energy to extract, consequently generating more pollution, and may be contaminated with heavy metals and radionuclides.

So the situation with phosphorus is similar to that of crude oil, where the problem is not so much "running out" of the resource as generating sufficient revenues from extraction and, at the same time, avoiding prices becoming so high that the demand for the resource is destroyed. In the present situation agricultural yields will be considerably affected if we are not prepared to face a decline in phosphorus availability. The real question, then, is when we will reach this condition. At present all predictive methods give uncertain results. With incomplete data on the reserve base, predictions are difficult.

A 2007 study showed that the Hubbert linearization method has been reliable for predicting future trends for specific regions or countries of production (notably Nauru and the United States).[34] However, the results, which by design are based on past production, are not as reliable for the whole world.[35] The 2007 analysis, for instance, did not include the accelerating production in China since 2006, which generated a new trend of increasing production. Based on about 40 years of production, the 2007 study concluded that the peak might have already occurred, in 1994, and projected an ultimate recoverable resource (URR) of 9,900 Mtons. However, if we apply the same approach used for the 2007 paper but focus only on the past 20 years of production data, the peak arrives in 2048 and the URR increases to 33,200 million metric tons. This projection is similar to another that used data on phosphate rock reserves and cumulative production between 1900 and 2007 from the USGS and the European Fertilizer Manufacturers Association; that study obtained a URR of around 3,212 million metric tons of elemental phosphate (about 24,100 million metric tons of bulk phosphate rock) and a projected peak in 2034.[36] Note that, in any case, even if we multiply the reserves by a factor of three to four, the peak moves forward only about 50 years into the future. At present not enough data are available for a final conclusion, but a regional versus global model may give a better picture of the future phosphorus production because it is able to identify countries or regions that are susceptible to grow, peak, or decline.

Sheikh Ahmed Zaki Yamani, former oil minister of Saudi Arabia, said some years ago, "The Stone Age didn't end because we ran out of stones." We can probably say that the phosphate rock age will not end because of lack of rock. It's much more likely that it will end because of a shortage of oil. Phosphate rock production depends on energy derived from oil for extraction, transformation, and transportation. It is difficult to imagine that we'll be able to maintain the present levels of phosphate production in the presence of a significant decline in oil production.

If we want to protect humanity from future starvation, we need to make substantial changes in the management and use of phosphorus. The problem here is one of time frame. While economic and political time frames are short, phosphorus production dynamics must be projected over several decades, with a management program envisioned over centuries. However, if these time frames can be managed, there are ways to counteract phosphate rock depletion.

Currently phosphates are inefficiently used, and there are many opportunities to improve efficiency. For example, phosphate present in the soil can be mobilized using relatively simple means like promoting mycorrhizal fungus growth through inoculation, no-till agriculture, and permaculture methods that rebuild soil naturally. These microorganisms can mineralize the soil's phosphorus to allow its easy absorption into plants. And even though phosphate rock has no substitute, it can be recycled. Returning human and animal manure to the soil is an essential part of phosphorus recirculation, as both these substances contain lots of phosphorus. The same holds true for green manures—cover crops that can be plowed into the soil to replenish its nutrients. Restoring healthy ecological connections between agricultural lands, forests, and water (oceans particularly) will be a fundamental element in the long term for conserving the Earth's fertility.

Long-Term Perspectives on Mineral Depletion

There may not exist an Achilles' heel for the mineral industry, but that doesn't prove that depletion is not a problem. It only means that we can adapt to the reduced availability of a single mineral commodity, or even of a few ones, by switching to different commodities. But we can't adapt so easily to the general depletion trend of all mineral commodities. To examine this point, we need to examine the whole extractive system. This approach is not new. It's precisely the one used by Donella Meadows, Dennis Meadows, Jorgen Randers, and William Behrens III, the MIT researchers who produced the *Limits to Growth* study in 1972, a study of concepts similar to those we have been discussing so far, but applied to the whole world's economy. The study ultimately described how exponential growth trends in population, pollution, industrialization, food production, and resource depletion could interact over the long term, eventually hitting up against limits imposed by a finite planet.

The story of *Limits* is so drenched in urban legends that it would take an entire chapter to unravel it, or even a whole book (which has been written[37]). Here, we can just mention the fact that almost all the legends still told today on the subject are just that: legends. It is not true that *Limits* predicted that the world would run out of some specific resources before the end of the 20th century. It is not true that it predicted imminent famines. It is not true that the study didn't use historical data as the basis for its models. It is not true that it was the result of an evil plot by multinational companies to exterminate inferior races and take over world government. In short, the study was not "wrong," as is commonly claimed still today.

That said, let's return to models. The *Limits* study was based on the same type of equations developed for the Lotka-Volterra model—that is, coupled differential equations. This kind of modeling received a great boost in the 1960s, when the development of digital computers permitted researchers to solve several coupled differential equations together. A pioneer in this arena was Jay Wright Forrester of MIT, who founded systems dynamics, a modeling method still used to simulate the complex interactions between forces in dynamic systems.[38] Forrester created the WORLD1 model to map global socioeconomic trends, and that model became the basis for the WORLD3 model used by the *Limits* researchers. World models are more complex than their ancestor, the Lotka-Volterra model, and include a much larger number of parameters.

The *Limits* study considered five main elements of the world system: nonrenewable resources (minerals), renewable resources (agriculture),

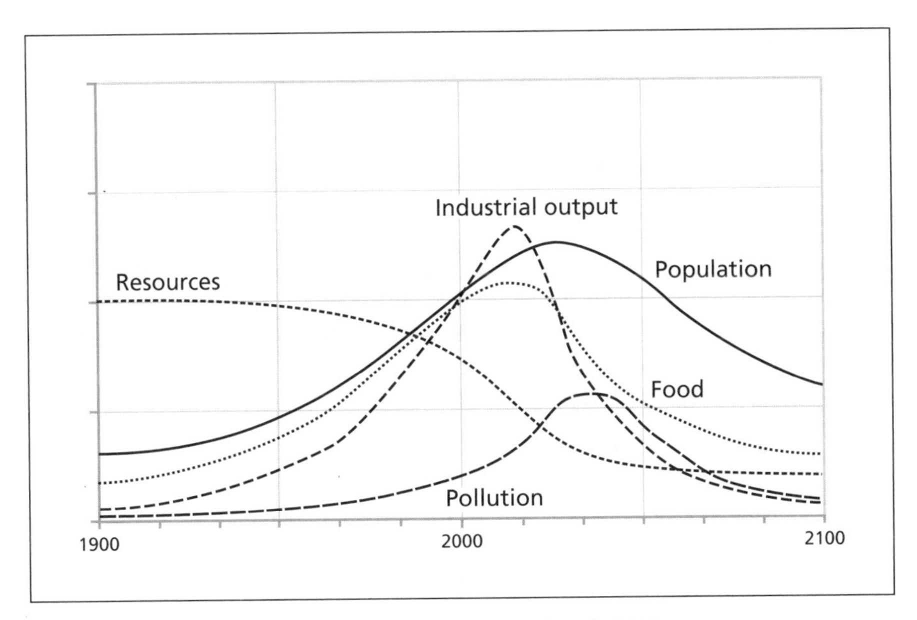

FIGURE 5.5. Base-case scenarios from the *Limits to Growth* study, 2004.

industrial capital, pollution, and human population. Using a complex set of data for each of these main drivers, they developed a number of scenarios with which to examine the future. In general, the *Limits* model showed that the world's economic system is affected by overshoot and that, because of the limitation in the available resources, it would not keep growing forever but would rather reach a peak in industrial and agricultural production at some moment in the 21st century and decline afterward. This productive decline would be followed by an irreversible decline of the human population. This behavior was most clear in the so-called base-case scenario, which used the most reliable data and made no assumption on the implementation of policies destined to avoid or mitigate overshoot. Figure 5.5 shows the calculations for this scenario as it appears in the 2004 version of the *Limits* study.[39] This recent result is not significantly different from the one that appeared in 1972. Recent appraisals have shown that the world system has followed the base-case scenario rather closely—a remarkably good result taking into account that these calculations were reported for the first time in 1972![40]

We can say that the *Limits* model leads to results similar to those obtained by the simpler Hubbert one—that is, a bell-shaped curve for industrial and agricultural production. The production of mineral resources does not explicitly appear in the figure, but it is bell-shaped, too. Unlike the curves of the Hubbert model, however, all the curves of the *Limits* scenario are skewed forward—that is, they show a decline faster than growth.

This fast decline, which we could define as collapse, generated by some mineral depletion models is a behavior that can be called the "Seneca effect," from the words of the Roman philosopher Lucius Annaeus Seneca, who noted in a letter to his friend Lucilius, "Increases are of sluggish growth, but the way to ruin is rapid."[41]

The "Seneca" behavior seems rather common in dynamic models that describe the exploitation of nonrenewable, or slowly renewable, resources. Only in extremely simple models that, for instance, do not take pollution into account is it possible to obtain the symmetric Hubbert curve. In all other cases the interaction of the elements of the system often leads to a rapid decline of the production curve. For instance, the decline may be generated by the effect of persistent pollution, which acts as a cost for the industrial system and thus reduces the amount of resources available for production. Alternatively, the system may allocate more resources to keep production ongoing, also in the form of technological improvements. This behavior leads to a higher amount of resources being extracted, but in so doing it exacerbates the depletion problem and causes the system to crash after reaching the production peak. This is a counterintuitive behavior typical of dynamic systems: actions that appear at first sight to solve the problem (increasing the efficiency of production) turn out to worsen it (generating a more rapid collapse).[42]

Even without the added trouble generated by the Seneca effect, the aggregated models based on system dynamics tell us that the decline in the world's overall production could start in the near future. And the economic crisis that started in 2008 could be the first hint that the decline is imminent.

It is still not clear whether the decline could be generated mostly as a result of resource depletion, or of pollution in the form of global warming generated by greenhouse gas emissions, or of some other factor. What we can say is that, although the future is not exactly predictable, the models give us the possibility of being prepared for it.

The history of mining has several dark sides, child labor among them. This boy worked deep inside a West Virginia coal mine in the early 1900s. Today, most nations have child labor laws, but children can still be found in some mining operations in Africa, South America, and Asia, according to the International Labour Organization. Mining continues to pose significant health and safety risks to workers and causes significant pollution as well, including pollution that leads to climate change.

6

The Dark Side of Mining:
Pollution and Climate Change

The story of Henry Russell, Scottish immigrant to the United States, tells us something of the harsh and dangerous world of miners as it was not long ago. In 1927, stuck at the bottom of a collapsed coal mine in Virginia, Russell took pencil to paper. He wouldn't come out alive from that mine, but we still have his last words to his wife: "How I love you, Mary."[1]

This is just one glimpse of a largely unknown world we should imagine as a place of misery and suffering. A British report by a mining commission in 1842 tells us about children commonly employed in coal mines for the simple reason that they were small and therefore could pass through smaller and less expensive tunnels. We read the story of an eight-year-old girl employed to open and close a gate inside a tunnel. She would stay in complete darkness most of the time, thanks to a certain perverse logic: a candle costs money, and what does the girl need it for? She was to keep it unlit, except for when her work was needed.[2] In 1910 Booker T. Washington, an American political leader who fought for the rights of blacks and of the poor, recorded the plight of Sicilian miners employed in sulfur mines. He tells us how Sicilian children were commonly used as workers—mistreated, beaten, and punished by burning their calves with miners' lamps.[3] Pictures of these Sicilian mines show that miners normally worked completely naked. It may have been because of the heat, but one can't avoid thinking that it was part of a strategy to humiliate them. Wrote Washington, "I am not prepared just now to say to what extent I believe in a physical hell in the next world, but a sulphur mine in Sicily is about the nearest thing to hell that I expect to see in this life."[4] That was the harsh reality of mining.

There are many more reports and stories from the world of mines that, today, can only make us shiver. To fight exploitation, miners often organized in unions; some were illegal and had to operate in secrecy. Such was the case for the 19th-century "Molly Maguires," a secret organization of Irish miners in the United States. In Britain the miners' unions were an important political force up to the 1980s. But they had been in decline for decades, with the

FIGURE 6.1. Sulfur has been mined in Sicily since ancient times. In 1900, Sicily had 730 sulfur mines, where some 38,000 miners toiled under inhumane conditions. By 1965 there were just 180 mines operating, and by 1983 only 13. Today, there is no more sulfur mining in Sicily.

parallel decline of British coal production. When oil production from the North Sea became an important economic factor in the British economy, the miners' unions became obsolete, a dinosaur of a bygone age. The clash of the miners against the Thatcher government in 1984–1985 ended with their complete defeat. In a way, they were defeated by crude oil. In general, neither unions nor secret societies appear to have been very effective in defending miners against exploitation.

In addition to the hard work, low pay, and various mistreatments (including physical punishments), mining was a dangerous job—possibly the most dangerous one in human history. Miners died because tunnels collapsed or gases emanated from the excavations. They died from incidents with explosives, and from all sorts of other failures and problems. The mines themselves were an unhealthy environment: they were often damp, and their air was oxygen poor, and at times laden with poisonous gases. The use of chemicals for extracting minerals created a whole host of new problems, as when mercury was used to recover gold from its ores. The miner's life was not just harsh but short. All told, mining has claimed an uncountable number of victims through its history. Eight million people are said to have died at just one single Spanish silver mining site—Cerro Rico ("rich mountain")—near Potosí, Bolivia.[5]

In more recent times, better technologies reduced the number of deaths among miners but couldn't eliminate them. The list of mining accidents during the past two centuries is long and detailed, with the US Office of Mine Safety and Health Research reporting almost 600 serious mining accidents in the United States from 1900 to 2010, with a total of almost 13,000 fatalities.[6] Of these, about 42 were major disasters with tens or hundreds of victims. Every country has a similar history of major disasters caused by collapses, explosions, or gas bursts. And those are only the incidents that ended with human casualties; they tell us nothing of the health problems that miners carry with them to old age, if they reach it.

Mining Debris

Until recently, the dark side of mining was something that affected only miners. For everyone else, the world of mines was far away and unknown—just as hell was said to be an ugly and terrible place but one from which nobody ever came back to give us a firsthand report. But that was gradually to change as the mining industry expanded to extract more and more materials. No man is an island, as John Donne said, and what was being done to the crust of our planet would, sooner or later, start affecting everyone. The first symptoms of trouble brewing started appearing in the second half of the 20th century, and at the beginning they seemed to be isolated and rare cases.

One of these cases occurred on the 21st of October 1966 in Aberfan, Wales. It was a rainy day like any other for the children of the school, but that morning rain would bring something much worse than just water. Around 9:15, when the children had just entered their classes, a dark wall of mud and coal crashed into the school. In a few minutes it was all over. The school had been buried under 12 meters of detritus arriving from a whole hill of coal waste accumulated from decades of excavations in the nearby coal mines. Among the victims were 116 children and 5 of their teachers.

The tragedy of Aberfan reminds us of an aspect of mining that we often forget: the fate of the extracted materials. Useful minerals are normally just a small fraction of what is extracted and processed from mines; the rest must go somewhere, and that becomes a problem when the volumes being extracted are large. As the volume being mined increases, the problem worsens, until it becomes serious—or even disastrous. In Aberfan the mining waste impacted human life directly, but debris generated by coal mining is an environmental problem of great importance that appears in all its evidence with the mining procedure called mountaintop removal. The process consists of blasting away entire mountains using dynamite in order to access the underlying coal seams or other mineral deposits. With the mountains go the forests and streams and wildlife that used to be there, and the rubble left in the process's wake smothers landscapes and waterways. The results can be described the same way Pliny the Elder described mining in Spanish Asturias in Roman times: *ruina montium,* "ruin of the mountains." In recent times, in the US Appalachian region alone, some 500 mountaintops have been removed and some 2,000 headwater streams have been buried or polluted in order to extract coal.[7]

Coal is just one of the mineral commodities that generate vast amounts of solid waste that must be disposed of. Copper is another example. Today we produce some 15 million tons of copper per year from minerals that contain it in a fraction of about 0.5 percent. It means that the total mass of rock extracted and processed is around 3 billion tons per year—larger than the total mass of

FIGURE 6.2. The remnants of the diamond mine of Mir, in Yakutia, eastern Siberia, Russia. Diamonds are mined in open pits and require vast excavations to recover relatively small amounts. The now-inactive Mir mine is so deep that the atmosphere above it has changed, and the surrounding airspace was closed to helicopter traffic after several were pulled downward by the airflow.

concrete produced every year in the world. To visualize this amount, imagine that you were asked to take care of the mining waste created by the copper contained inside your new car. An average car contains about 50 kilograms of copper, mainly in the form of wiring. So, on your way home from the dealer, you would be followed by a truck that would then proceed to dump about 1 ton of rock in front of your door.

Like the landscape of the Appalachian Mountains, which is being deeply changed by mountaintop removal, areas around the world are being deeply changed by the mining of all sorts of mineral commodities. If we were to take into account the waste generated by all mining activities, we could easily arrive at levels of several tens of billions of tons of rock being processed and dumped somewhere. And if the gradual depletion of mineral resources were to lead us to mine from less concentrated ores, that would bring us even larger amounts of solid waste.

In addition to creating a monumental waste disposal problem, mining often leads to gigantic holes in the ground that are left to fill up, gradually, with water and debris. Some are so large, such as the giant pit of the Mir diamond mine (see figure 6.2) in Russia, that will remain visible for centuries, perhaps millennia.

But mining doesn't always consist of just digging up inert dirt and rock from holes in the ground. Often it involves using chemicals and reactive substances of all kinds. Several areas of central California are still contaminated today by the mercury that miners used to extract gold during the 1849 gold rush.[8] In modern times we have developed a variety of methods that use poisonous and polluting chemicals to recover useful materials from the ground. For instance, cyanide is used to extract gold from ores, a process that makes low-grade ores useful but creates tremendous environmental damage. A combination of chemicals involving hydrogen peroxide and sulfuric acid is used for on-site leaching at uranium mines.[9] There are other cases of mining by in situ leaching, including for copper, and probably more will be developed in the future.

Mining activities can also directly create pollution in the form of noxious liquids or gases, especially in the case of fossil fuels, for which mining operations are rarely completely clean. A recent case is that of "fracking," a method of injecting liquid solutions into the ground to fracture the rock and facilitate extraction of crude oil and natural gas. The fracking liquids are often laced with acids, solvents, and other chemicals that may contaminate water sources, and the process may generate earthquakes.[10] (See "Fracking: The Boom and Its Consequences," page 55.) Even conventional oil drilling often generates waste products in the form of methane vented in the air or burned in place. And speaking of burning in place: consider also the case of underground coal fires, or "coal seam fires," which can occur in underground coal seams, in seams that have been exposed, and in aboveground coal stockpiles or waste deposits. These fires can smolder for decades, causing considerable underground volume loss (which can lead to surface collapse) while also sending massive amounts of toxic gases into the atmosphere.[11]

Waste, Waste Everywhere!

The pollution directly created by mining is just the first link in a chain of ever-growing pollution. Once extracted, minerals are processed and transformed into marketable products. These products are then "consumed"—that is, destroyed and discarded. All the stages of the process generate waste: manufacturing industrial products generates industrial waste, while consumption of the products generates urban waste. Where does all this waste go? The law of the conservation of mass says that it must go somewhere, and that is a serious problem. Waste is the ultimate product of mining.

To do damage, the products of mining don't have to be poisonous or reactive; in some cases sheer volume is enough to create damage. One example is concrete, a building material used everywhere in the world. We don't usually consider buildings and other structures made with concrete to be waste, but they do tend to decay, albeit slowly, and—new or decayed—the large area they occupy is becoming a problem.

We don't have precise data on the fraction of the world's land surface now covered with concrete in the form of roads, parking lots, buildings, commercial centers, and all the rest. However, recent studies are starting to converge on reasonably consistent values ranging from about 0.5 percent to about 3 percent.[12] Translated into area, these values correspond to a minimum of 700,000 square kilometers and to a maximum of about 3 million square kilometers. To aid in visualizing these areas, consider that the first figure compares to France (550,000 square kilometers) and the second to India (3.2 million square kilometers).[13] No matter which result we consider more reliable, it is a huge area, and the data show that such building takes place mostly in flat and fertile areas. There, the fraction covered by human-made structures is much larger than the world average. For instance, recent data for Europe indicate that, in 2011, the most urbanized European states were Holland and Belgium, which had, respectively, 13.2 percent and 9.8 percent of their surface covered with permanent structures.[14] Most of these flat areas could have been used for agriculture.

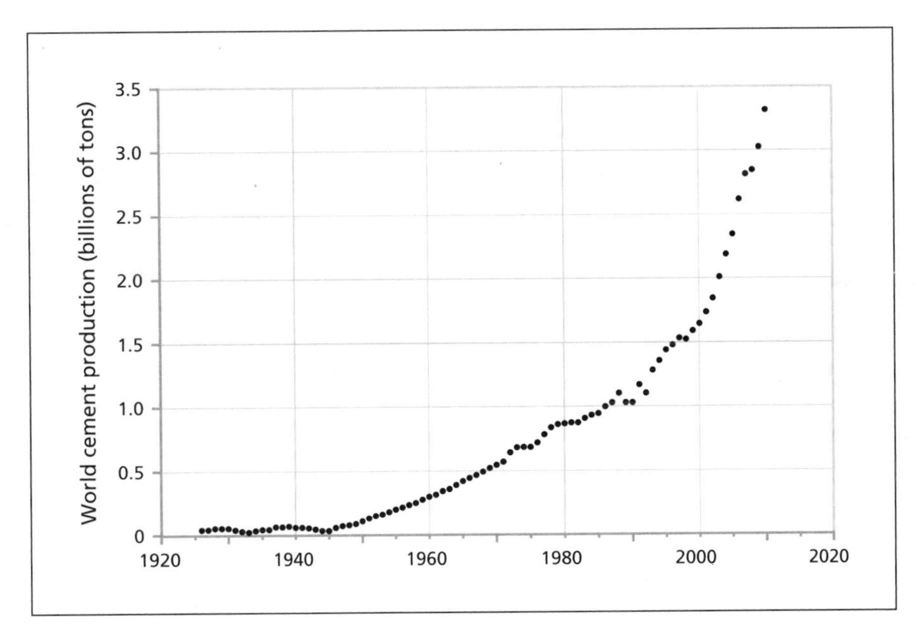

FIGURE 6.3. World production of cement. Cement and construction materials are probably the fastest growing mineral commodities in the world, in terms of the amounts of materials they require to be mined.

We have no data on how fast has been the growth of paved areas, but if it has been proportional to the quantity of cement produced, it has been spectacular. Perhaps we could call this explosion of paved and built areas the "Trantor effect," referring to the fictional planet Trantor, capital of the Galactic Empire invented by Isaac Asimov for his Galactic Empire trilogy.[15] In Asimov's story Trantor is described as a planet of 40 billion inhabitants whose land surface is 100 percent covered with human-built structures. It is a fictional place, of course, but if we extrapolate the rate of increase of paved land on Earth we see that, theoretically, we could arrive at something similar in not much more than a century. That will never happen, of course, but the insistence on paving as much land as possible, as fast as possible—which seals the soil from water and makes it inaccessible for photosynthesis—could start giving us serious problems much before we transform Earth into a twin of Trantor.

Ordinary waste doesn't occupy the same enormous areas and volumes associated with concrete, but it creates a problem nevertheless, and often we don't fully understand its size. The problem is something relatively recent in human history; a few centuries ago people could never have conceived of the existence of a "waste problem." But now it has become so big that it already seems to be intractable. Waste is mostly the result of mining materials that went through the industrial cycle and were discarded afterward. We still have a lot of material to extract, and a lot of the material that has been extracted is now somewhere within the industrial system or in use in people's homes, commercial buildings, and the like. An enormous amount of material has not yet become waste, at least officially, but it will during the coming decades and centuries. This great mass of waste in waiting will make even worse the problems that we are facing nowadays.

The problem with waste is not so much related to the space it occupies; we won't anytime soon begin demolishing homes to make space for landfills. The real problem is pollution, a term that can be applied to any substance that has an active damaging effect. We have, with all our mining, extracted from the Earth's crust substances that had been buried for millions—or even billions—of years and that caused no problems to anything or anybody as long as they remained underground. But once extracted, processed, and concentrated in new chemical forms, these substances—never before seen in the modern global ecosystem—began creating a series of problems related to human health and to the general health of the ecosystem.

The problem is especially evident with radioactive waste. Many of today's radioactive elements never existed on the Earth's surface before humans began mining or creating them, and as a consequence living beings have evolved no mechanisms for dealing with them. Consider plutonium as an example. Apart from a brief appearance as the result of a natural nuclear reactor created by geological processes some 2.5 billion years ago, plutonium has never existed in measurable amounts in the Earth's crust. But it exists today as a by-product of the operation of human-built nuclear reactors. Biological organisms never dealt with plutonium, and it is not surprising that it turns out to be one of the most poisonous substances known.

Getting rid of plutonium is not easy. The isotope 249 of plutonium has a half-life of 24,000 years. This means not that it will disappear in that time span, but that in 24,000 years 50 percent of the plutonium that we have created so far (and are still creating) will still exist. In 100,000 years, 6 percent of this plutonium will still exist. The only way to engineer its disappearance over shorter time frames is to burn it in reactors, but that generates further and different radioactive isotopes, which create other problems. Perhaps we could shoot it to space or to the moon, but that would be terribly expensive.

The only way we have conceived so far to get rid of long-life radioactive waste is to bury it in areas of the crust that we believe will be stable for long times. Yet this creates new features in the Earth's crust: new forms of high-concentration radioactive mineral deposits that are the result not of geological activity but of human industrial activity. How our descendants will deal with this problem is impossible to determine. From an ethical point of view, we are doing them a tremendous disservice. We are passing onto them heavy loads of dangerous materials, and it is not at all obvious that they'll have the scientific and technological tools to deal with the problem, or even that they will be able to recognize that it exists.

If managing radioactive waste is an extremely difficult problem, at least radioactive materials are created in small amounts and are easily traceable because of their radioactivity. A more general problem exists with heavy metals. Many of these are toxic, and all are alien to the ecosystem in the quantities in which they are being created and dispersed today. Once they complete their cycle through the industrial system, the fate of these metals may involve disposal in landfills. In some sense, this is a way to place them back where they came from: underground. Provided that the landfill is well built and doesn't leak to the water aquifers, we can expect that the materials thrown in there today will not leach away for a long time.[16] In this context "a long time" means "some centuries," but that doesn't mean the problem will not be faced sometime in a remote future by our descendants. Again, we are putting a heavy burden on them.

But not all landfills are built with the idea of lasting for centuries, and there have been many cases of landfill leakage with tragic consequences. Perhaps the best-known example is the Love Canal landfill in New York State. A chemical company had buried tons of toxic waste there, and in later years housing and other urban developments were constructed on the site. In the 1970s and 1980s it was discovered that the landfill had been leaking toxic waste, generating a host of serious repercussions in the local population, from nervous disorders and cancer to birth defects.[17] Several similar cases exist, where chemical companies released toxic waste that later affected the local community near the plants. In many cases, however, the chemical industry has exported its toxic waste to remote areas and dumped it there without much attention to the health and safety of the residents—often poor and without the capacity to oppose this kind of action. This kind of activity is strongly suspected in areas surrounding the Italian city of Naples,[18] but we have little quantitative data on the phenomenon, which is likely to be widespread in the world. Its long-term consequences are still to be discovered.

Even beyond dumps and landfills, legal or not, heavy metals are being dispersed all over the world in the form of fine particulate and volatile compounds that can be inhaled or eaten. In some cases dispersed particulate is the result of incineration. Theoretically, a modern incinerator is equipped with filters that greatly reduce the amount of powder emitted at the smokestack. But no filter is 100 percent efficient, and the smaller a particle is, the less efficient a filter is at trapping it. Across the world, nano-sized particles, suspected to be the most damaging kind for our health, make their way from smokestacks.[19] Regulations and monitoring, as they are implemented nowadays, may be largely insufficient to stop the problem. Even though considerable efforts are being made to try to manage the problem, not all incinerators in the world have good filters, and many have no filters at all, especially in poor countries.[20]

Nevertheless, we seem to be fixated on the idea of using expensive and polluting incineration to deal with our waste. Maybe our fascination with incineration has something to do with the ancient fascination that humans have with fire, something that comes all the way from Paleolithic times. Unfortunately, however, incineration only gives us the illusion of getting rid of waste; instead, it transforms it into compounds that are often more dangerous and more difficult to deal with than the original ones.

On the whole, however, incineration produces only a modest contribution to the amount of dangerous particles present in the ecosystem. More important sources are the result of industrial combustion processes, with one of the most important being coal burning. Normally coal contains traces of heavy metals (including radioactive ones) that, when burned, are emitted in the atmosphere in the form of small particles. Filters at the top of the smokestack can remove most of these particles; this is one of the characteristics that give rise to the very optimistic concept of "clean coal." But, again, most coal plants in the world don't have filters, and even when they are present, filters cannot eliminate all the particles that are produced, especially the very small ones. Independently of combustion, metals can be transformed into powders as the result of abrasion, corrosion, and other industrial processes. These are unavoidable processes that affect most metallic objects. The result is that large amounts of heavy metals are routinely dispersed in the environment.

So the waste problem is not only one we don't know how to solve, but also one that is becoming larger as the products of mining pass through the economy—a retardant effect whose size we are not yet equipped to appreciate. Sure, some say that waste is not really a problem but an opportunity. If we could recycle it efficiently, we could turn waste into resources and feed the economy with what we recover from landfills. This is surely a possibility, but it is subject to the same problem of the "universal mining machine" that we discussed before. Useful minerals are dispersed in waste in low concentrations, and their recovery is expensive in terms of both energy and money. We will examine the problems and opportunities involved in waste recycling in more detail in the next chapter.

Heavy Metal Waste: Mercury and More

To get some idea of the size of the heavy metal pollution problem, let's consider a specific example: mercury, one of the most toxic metals known. Mercury is very rare in the Earth's crust; its average concentration is about 80 parts per billion (ppb), much smaller than that of most heavy metals. Before the industrial age, extremely small traces of mercury were present in the ecosystem, mainly as the result of volcanic eruptions. But human mining activity extracted mercury from concentrated deposits, mainly in the form of cinnabar (HgS). Now large amounts of mercury have been dispersed all over the planet.

Today we recognize the high toxicity of mercury and often legislate it away from industrial production. It is possible that, at some moment in a not-too-remote future, the mineral production of mercury metal will dwindle to zero. That would be a good thing for human health, but a lot of mercury has already been extracted and remains somewhere around us. We can estimate from the production data that the total mercury produced from mines up to now is on the order of 500,000 tons. But this is not the only source of mercury resulting from human activity. Coal extraction and combustion in coal plants generate an amount on the order of 1,500 tons per year.[21] It is difficult to estimate how much mercury this process has dispersed in the atmosphere over the years, but it is probably on the order of a few hundred thousand tons, which can be added to the mercury produced from mines. Where has all this mercury gone?

In part, it is still around in the various objects and devices that use it: thermometers, fluorescent lamps, batteries, and dental fillings, to name a few. The data available on these amounts are highly uncertain and probably underestimated.[22] It appears, however, that the mercury still present in the industrial system is possibly around 50,000 tons. That would represent about 10 percent of the mercury extracted from mines. The rest has followed the destiny of urban and industrial waste, being landfilled, incinerated, or simply dumped somewhere. A large part of this mercury is by now embedded in the ecosystem. According to the available data, some 200,000 tons of mercury are present in the first 15 centimeters of soil.[23] More is present in the oceans as dispersed powder or soluble compounds.

Even with these data, we see that we can account for only a fraction of the total. We don't know where the missing mercury could be; maybe it is in landfills, or maybe it is in forms that we have difficulty detecting and estimating. In any case, the half-life of mercury in the ecosystem is about 3,000 years. So even if we were to stop mercury production today, the amounts already produced would remain with us for thousands of years.

In the meantime, we are in contact with this dispersed mercury every day. We accumulate it in our bodies by breathing, drinking, and eating. The several thousand tons of dispersed mercury present in the seas are absorbed by living creatures and accumulated in progressively larger amounts as we go up the food chain—a process called "bioaccumulation" that concentrates mercury in the body of top predators. As humans are at the very top of the food chain, we are probably the species most at risk from mercury accumulation, which is perhaps what we deserve, as we are the ones who have generated it.

Mercury is a neurotoxin, being damaging to the nervous system, but it also damages the liver, negatively affects fertility, and more. It is especially damaging in its volatile liquid form, known as dimethylmercury; a single drop on human skin is enough to kill. That was the destiny of Karen Wetterhahn, professor of chemistry at Dartmouth College in New Hampshire.[24] In 1996, as she was handling dimethylmercury while performing experiments in her laboratory, a drop of it fell on her hand. Despite the glove that she was wearing, that drop was enough to kill her in a slow and painful agony that lasted several months. Other cases have involved multiple victims, such as in the Japanese village of Minamata, where in the 1950s tens of thousands of people were poisoned by eating fish caught in an area of the sea that had been contaminated by the mercury released by a chemical plant industry.[25]

Every one of us, unavoidably, has some mercury in our body, normally in quantities deemed "safe." However, the Web is full of reports of people who claim to have been cured of various symptoms, from headaches to skin rashes, by removing dental fillings made of "amalgam." A typical amalgam filling is about 50 percent mercury. There are no reliable epidemiological data on these cases, and the amount of mercury absorbed by the body from amalgam is believed to be low.[26] But these reports can't be ignored, either.

What can we expect for the future? Mercury is still extracted from mines and released into the environment by industrial processing and coal burning. Hence we are probably going to see an increase of its average concentration in the human body. It is impossible to say how this will affect us; we simply have no data on the long-range effects of small amounts of mercury on human health. Some recent studies show that the great Permian extinction of 250 million years ago was associated with high levels of mercury resulting from volcanic eruptions.[27] Probably it was not this mercury that caused the extinction; still, it is an uncomfortable fact. Today, we can only wait for the results of this great planetary experiment we are conducting on ourselves.

Mercury is just one of the minerals for which we've made ourselves into human guinea pigs. The US Agency for Toxic Substances and Disease Registry (ATSDR) keeps a record of the various toxic substances generated by mining and industrial activities.[28] The most toxic metals on the list are arsenic, lead, mercury, and cadmium. For these metals, the situation is similar to that of mercury: anthropogenic emissions largely surpass natural ones, and little is known about the long-term effects on human health. Even metals that are relatively common in nature may be transformed by industrial processing into forms that are highly toxic. For instance, chromium in the form chromium-3 is relatively common in the Earth's crust and is also a necessary element for human metabolism. But there exists another chemical form of the element, known as chromium-6, that is extremely rare in nature but commonly used by the galvanic industry for chromium plating. This form of chromium is highly carcinogenic and poses serious pollution problems.

The problems generated by single substances are compounded and amplified by their combinations. We are not exposed to chemicals one at a time and for limited spans of time—as happens in controlled studies—but in combinations of tens or even hundreds of them, continuously, in our daily lives. Just think: the number of chemical substances registered for industrial use is approximately 100,000 in the European Union and 84,000 in the United States. We inhale artificial chemicals, eat them, and drink them all the time: It is like one of those "surprise" cocktails that you can order without knowing the ingredients. We just don't know what the cocktail will do to us in the long term.

To make the picture even more complex, chemicals often break down in the environment and generate new substances with unknown properties and effects. Even if we were to know exactly what we are exposed to and in what amounts, the task of obtaining reliable data on how these exposures impact human health would be an impossible one. It is already a slow, difficult, and expensive task to study the effect of single substances, but when examining exposure to multiple substances, the task becomes monumental.

It is often said that exposure to chemicals is not a problem because the average life span in the Western world keeps increasing. That's true for some countries, but not for all of them, and in any case, we don't have a "control experiment" that could tell us what our life expectancy could be in the absence of these human-generated chemicals. And how about quality of life in a world where we are more and more dependent on medical care to be kept alive?

Some data seem to indicate that, indeed, the healthy life expectancy–the expected average number of disability-free years—may have been declining in some European countries.[29] But measuring these quantities is difficult, since the concept of quality of life, just as that of health, is mainly based on people's individual perception. At present, we can only say that we are performing a giant experiment on how human health is affected by small doses of heavy metals and other human-generated chemicals. The results will only be clear in the future; for the time being, we are the guinea pigs.

Waste as Greenhouse Gases

The problems with heavy metals generated by mining are already very serious, but at least they don't seem to be directly affecting the homeostasis of the planetary ecosystem. That cannot be said of another experiment we are carrying out that has the potential to cause even larger damage. It is the emission of greenhouse gases, principally in the form of carbon dioxide (CO_2).

The combustion of fossil fuels is the major factor generating the observed rise of CO_2 concentration in the atmosphere. Of the various forms that these fuels can take (coal, gas, and oil), coal is the most damaging in terms of climate change; that is, it is the fuel that generates the largest amounts of CO_2 for the same quantity of energy generated. So coal, we see, has dramatic impacts all around. Even so, some hold high hopes that coal can fuel the economy when oil and gas decline—a subject explored in "Peak Coal."

The concentration of CO_2 in the atmosphere has increased from about 270 parts per million (ppm) in the preindustrial era to the present (2013) values of 400 ppm, and those concentrations keep growing.[30] Of the 300 billion tons of carbon generated by the combustion of fossil fuels, about half is still present in the atmosphere, as detected by isotope studies; the rest is mainly dissolved in the oceans.

The effects of the sudden increase in CO_2 concentration we're experiencing today in the atmosphere are still mainly to be seen, but we know something of what to expect. It is often said that climate science is a question of models, with the corollary that models are uncertain and that for this reason we cannot be really sure of how warm our planet will become. Sometimes it is said that it is not even certain that it will warm at all. This is a bad misunderstanding of modern climate science. If it is true that models are an important part of the field, it is also true that with more data being collected and more studies being performed, the field called paleoclimatology is becoming more and

Peak Coal

Werner Zittel and Jörg Schindler

Conventional wisdom would have us believe that coal is abundant, cheap, and able to meet our needs for several hundred years. As a consequence, some say that it will be possible to replace the dwindling supply of crude oil with synthetic fuels created by turning coal into liquid fuels. How realistic are these hopes? Not very, it seems, even without considering that a major shift toward coal-derived fuels would have dramatic climate impacts. The real problem may be that coal is not as abundant as some optimistic assessments tell us.

As far back as 1865, William Stanley Jevons applied what we'd today call a "systems approach" to assessing the future availability of coal by analyzing geological, technological, economic, and even ecological factors. He concluded that, because of the finiteness of the resource and the increasing costs of extraction, future production would follow a steadily growing curve up to a maximum; then it would be followed by a steady decline. In other words, he outlined the concept of "peak coal," though he did not use that term. The basic concept of peak coal has been largely forgotten, and it almost never appears in modern forecasts. However, Jevons's work remains the basis for analyzing the future production trends of all mineral energy sources.

Today coal is used in one form or another across many industries, but its main uses are in power stations, cement production, and the iron and steel industries. It occurs in many different qualities, with many different classifications. But the international coal trade frequently classifies it, according to its use, in two categories: thermal coal (also known as steam coal or non-coking coal) and coking coal.

There is no global standard for assessing reserves (known extractable deposits) or resources, which is one reason the available data on coal reserves and resources are often poor quality. Other reasons include the fact that some countries, including China and the former Soviet Union, report obsolete reserve data. According to a 2013 World Energy Council (WEC) report, in fact, in that year 67 countries submitted data unchanged from previous years, while 11 submitted revised data and 9 countries (Armenia, Bangladesh, Georgia, Laos, Mongolia, Tajikistan, Belarus, Bosnia-Herzegovina, and Macedonia) reported reserves for the first time.[31] The most recent data for Afghanistan and Vietnam are from 1965. The countries with unchanged reporting cover 89 percent of the total.

Arbitrary revisions of reserve data, leading in some cases to dramatically lower values, also present obstacles to accurate projections. For instance, in 2004 proven German hard coal reserves were downgraded by 99 percent. Between 1993 and 2010 Poland downgraded its hard coal reserves by 85 percent and its lignite (or brown coal) reserves by 90 percent. Further significant downgrading was reported for the UK, South Africa, Kazakhstan, the Czech Republic, and Hungary. The same was true of India, though that downgrade was caused by data transmission errors between national and international agencies.

In some countries resources have been considerably upgraded, but at the same time reserves have been downgraded and production has reached historical lows. For instance, at the end of 2010 Japan's hard coal resources were stated at 13.5 billion metric tons and reserves at 340 million metric tons. Despite the theoretically huge resources, production in Japan declined to an insignificant 0.9 million metric tons per year, while the country imported 186 million metric tons in 2010 (it was the biggest coal importer in that year, only to be overtaken by China in 2011). Similar inconsistencies can be shown for many other countries.

It appears that there is no feasible way to determine how reliable the reported data are, but some evidence—for example, data compiled by the WEC from 1987 to 2010—suggest that the presumption of abundant and cheap coal in the long term is not justified. In this period proven global coal reserves were downgraded by 739 billion metric tons (46 percent), from 1,600 billion metric tons to 861 billion metric tons. Cumulative coal production amounted to 123 billion metric tons, causing reserves to be downgraded by an additional 616 billion metric tons from 1987 estimates. Specifically, the downgrading of reserves contradicts the conventional wisdom that with rising coal prices and technological advances, coal resources will be converted to reserves. In reality, the reserves-to-production (R/P) ratio has fallen, bringing supply estimates from 400 years in 1987 to under 120 years in 2011, and a net conversion of resources into reserves has not been observed in the last two decades.

Projections of future coal supply can only be based on reserve data. The question is, what are the producible volumes and production profiles over time? Many limiting factors in specific regions can reduce producible volumes to amounts far below geologically determined reserves. Such factors can involve technical, economic, legal, or environmental issues, or the attitudes of citizens affected by mining-derived pollution. Therefore, for all regions where detailed regional analyses have been carried out, a lower reserve value than the WEC value should be used for modeling future supply. Only by taking into account non-geological factors can we derive adequate reserve figures.[32]

This approach has been confirmed by detailed case studies of regions with big coal deposits in the United States. One example is the Gillette coal field in Wyoming: reserves are stated at 192 billion tons, but after considering all existing limitations, approximately only 70 billion tons are technically recoverable and only about 9 billion tons are economically extractable—less than 5 percent of the geological reserves.[33]

Peaking of coal production is already a reality in many regions, including Japan. In Europe major countries peaked long ago: the United Kingdom in 1913, Germany in 1958, France in 1973, and Poland around 1990. Hard coal production in Europe as a whole peaked around 1960 and today is phasing out. Is Europe a paradigm for the rest of the world?

Despite the uncertainties, it is still possible to get some idea on the relative productive potentials of different regions in the world, as shown in table 6.1.

The table lists data for the four major world producers (the United States, Russia, China, and Australia), which together hold almost 60 percent of the world's reserves. Coal production, unlike that of oil, is concentrated in a small number of countries, and we need to add only four more to the list above (Kazakhstan, Ukraine, India, and South Africa) to define nearly 95 percent of global hard coal reserves. Future production in these countries will determine the global production.

China is by far the largest coal-producing country in the world, but the situation with reserves is rather unclear. The most recent reserve

Table 6.1. Most Important Coal Countries in 2012

	Rank 1	Rank 2	Rank 3	Rank 4	Biggest 4 share of total (%)
Reserves [Mtoe]	USA 133,000	Russia 74,000	China 64,000	Australia 42,000	55
Production [Mtoe]	China 1,825	USA 516	Australia 241	Indonesia 237	73
Net Exports [Mt] (Mtoe)	Australia 316 190	Indonesia 304 182	USA 105 63	Russia 97 58	74

Note: Mt = million metric tons; Mtoe = million metric tons of oil equivalent (1 toe = 6,841 barrels oil; 1 toe = 0.6 metric tons coal with an energy content of 25 MJ/kg).

numbers for China date from 1992. The corresponding reserve to production (R/P) ratio would yield a supply of 31 years. But cumulative production since 1992 amounts to about 30 percent of these reserves, and the R/P ratio corrected by the cumulative production is just 20 years. Even though the reliability of these data is low, it is likely that production growth will slow down in the coming years and that peak production will be reached in 5 to 10 years.

The United States is the second biggest coal producer and holds the largest declared reserves, comprising about 30 percent of the global reserves. US coal production has nearly doubled since 1970, but the growth rate has declined since the mid-1990s and from 2008 production has reached a plateau and shows signs of decline. Note also that the production of high-quality coal peaked around 1990 in the United States—only the production of lower-quality coal has increased. The energy content of US coal production peaked in 1998.

Detailed analyses of coal deposits in the United States show that the reported reserves are often overstated. Probably half of these reserves will never be produced due to various substantial obstacles. A sizable growth of US coal production could only be achieved in Montana, where there are huge deposits of subbituminous coal, although with low energy content. In theory, the available reserves in Montana would enable a growth of the US coal production by about 50 percent. However, in view of the economic and legal conditions in Montana this is totally unrealistic. Therefore it is very likely that US coal production has peaked already. The remarkable rise of US exports since 2009 is a consequence of reduced domestic coal consumption.

In Russia, an expansion of coal production is possible, but the major part of Russia's not-yet-developed reserves is in Siberia, in areas lacking any transport infrastructure. So while an expansion of Russian coal production is likely, it will probably be at a small scale that cannot be exactly quantified at the moment.

The available data can be used to build up a model of future coal production worldwide based on a detailed analysis of each producing country.[34]

As the scenarios in figure 6.4 suggest, and starting from the assumption that global coal reserves will not grow substantially in the coming two decades, we can conclude:

- With a very high probability, the global peak of hard coal production will be reached before 2050, probably even as early as 2020.
- The level of peak production will be mainly determined by the future production rate of China. World peak production will probably range between 8 billion and 10 billion metric tons (compared with 7.86 billion metric tons in 2011).
- The level of global peak production will be dependent only to a small degree on reserves in Australia, Russia, and the United States because of the very long lead times for their development.
- These scenarios do not account for possible future restrictions from climate policies. Economic aspects are not explicitly addressed but are implicit in the historic production development and in the downgrading of WEC reserve data.

In addition, the fact that only about 15 percent of the global coal production was traded internationally in 2012 shows us that the countries owning the biggest reserves have only a limited export potential. In 2013 China and India together imported 364 million metric tons, twice as much as the imports of Japan in 2010 when Japan was still the biggest importer worldwide. Only 10 years ago China was exporting 70 million metric tons.

This rapid change in international coal trade resulted in a doubling of the global import/export market since 2001. The additional

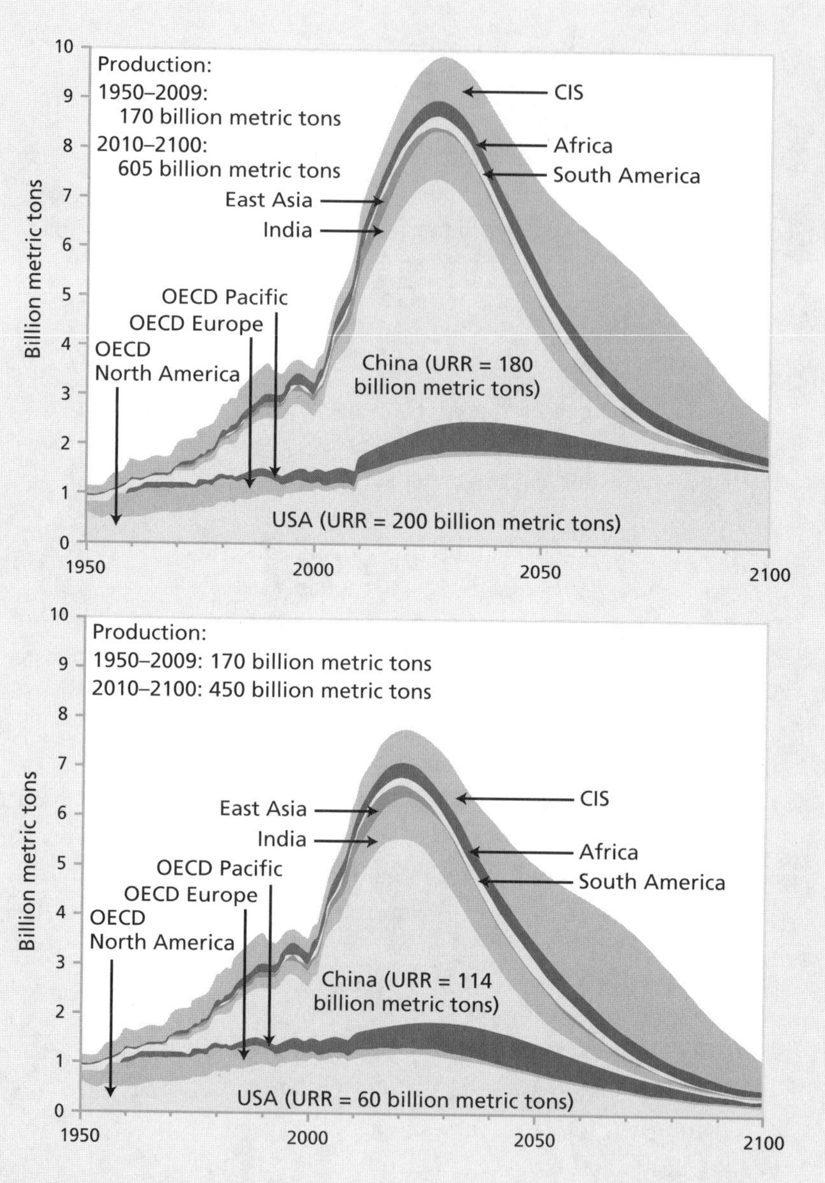

FIGURE 6.4. Scenarios of global hard coal supply, 1950 to 2100.[35] The top scenario is based mostly on the WEC reserve data (except for China, where bigger reserves than stated by the Chinese authorities have been assumed). In this scenario, of the global hard coal reserves of 675 billion metric tons in 2008, 615 billion metric tons will be produced from 2009 to 2100. The scenario below—the more likely scenario—is derived by downgrading some of the doubtful WEC reserve data.

volumes were mainly supplied by Indonesia, which expanded its coal production. Yet it is foreseeable that Indonesia will peak within the next five years and its exports will subsequently decline. Since exports from South Africa in 2012 were similar to exports in 2005, the future gap in export capacity can probably be closed only by increased exports from countries of the former Soviet Union, Colombia, and Australia. Since the coal demand in China and India will continue to grow, it is likely that the near future will see shortages and rising prices on the world market.

So cheap and abundant coal for decades to come is not a very likely scenario. Rather, we will see some further growth in global coal production followed by a peak in the not-so-distant future. Therefore, there is definitely no scope for substituting for oil and gas with coal. The peak of all fossil fuels is already in sight.

more important for the understanding of what is happening and what we may expect to happen in the future.

From paleoclimatology, we know that the concentration of greenhouse gases, mainly CO_2, in the atmosphere is not an isolated factor that changes temperatures according to a single sensitivity factor, whose exact value we still don't know. It is part of the carbon cycle and affects the whole Earth system. The carbon cycle, and variations in its greenhouse effect, has been an integral part of the ecosystem since remote geological times. Its importance can hardly be overestimated. It is said that nothing makes sense in biology without the concept of evolution, and we could say that nothing makes sense in climatology without the concept of carbon cycle and of the role of CO_2 as a greenhouse gas.

We have already seen in previous chapters how the evolution of the Earth's ecosphere has been deeply affected by changes in CO_2 concentrations, which have maintained the planetary temperature within the boundaries needed to maintain liquid water at the surface. The present question is what is going to happen as a result of the present perturbation caused by the burning of a large fraction of the carbon compounds that were buried by bio-geological processes over past geological ages. What is especially impressive, here, is how fast the removal of this carbon has happened in comparison to its burial time. The hydrocarbon deposits existing today likely took at least several hundred

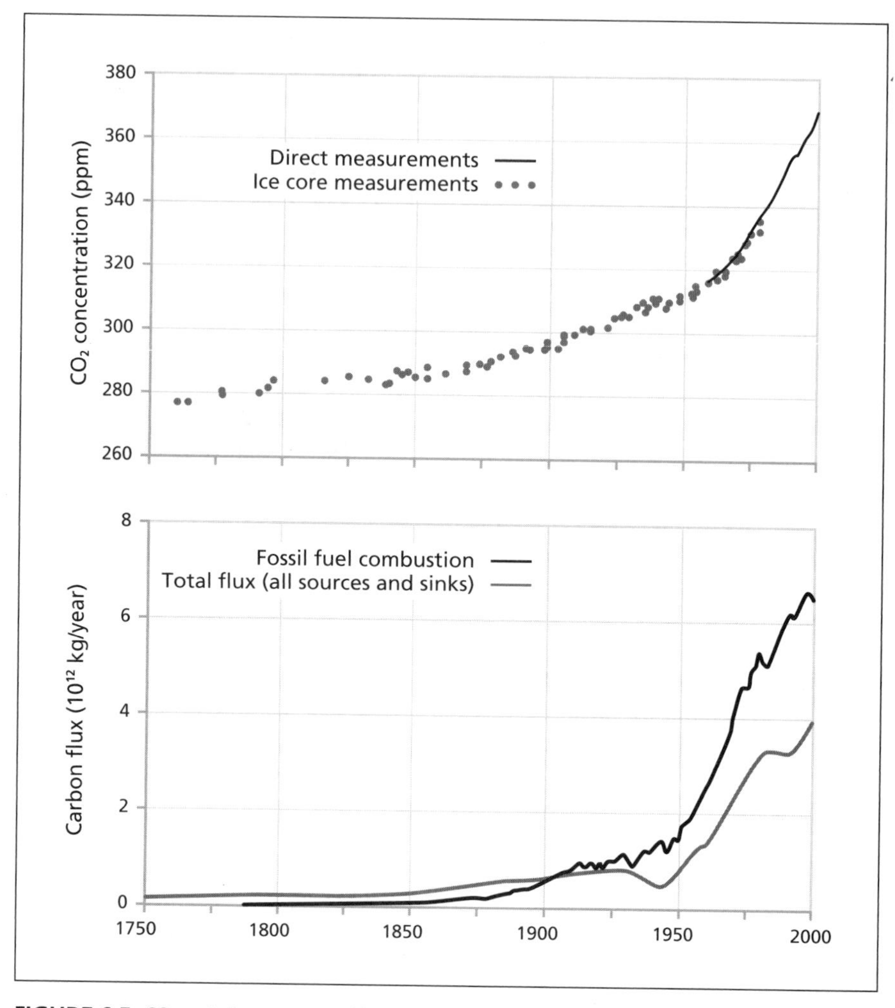

FIGURE 6.5. CO_2 emissions generated by the combustion of fossil fuels. The increase in the CO_2 concentration in the atmosphere seems to be unstoppable, and its warming effect may cause us tremendous damage in terms of climate change.

million years to form, but we are extracting them and transforming them into CO_2 in a timespan that may sum up to just a few hundreds of years. Again, the present events can be seen in the context of paleoclimatic data, which show us that the present burst of CO_2 emission appears to be more intense and more rapid than past ones.[36] If we consider that past CO_2 events were associated with major planetary disasters and mass extinctions, we can understand the gravity of the problem we are facing.

The problem of CO_2 concentration increase could be compounded and enhanced by secondary effects, such as the release of methane (a more powerful greenhouse gas than CO_2) currently locked in permafrost areas and at the bottom of oceans in the form of hydrates.[37] These hydrates are a true climate bomb that could lead to a catastrophic increase in the planetary temperature. According to some studies, as the result of runaway warming, some areas of the Earth could become too hot for humans to survive.[38] But even without arriving at such extremes, human survival would be indirectly threatened by the disastrous effects that a change of this magnitude could cause to agriculture.

The recent (2012) near complete melting of the North Pole's ice cap is an indication that global warming not only is occurring but may be accelerating, probably as the result of feedback phenomena involving methane or other factors. In 2013 the North Pole recovered some ice coverage in comparison to 2012, but on average the melting trend continues unabated. From the available data, we expect not only a considerable warming of the whole planet, but also a very rapid one, at least in geologic terms.[39] In this sense the scenarios produced by the Intergovernmental Panel on Climate Change (IPCC) could be considered optimistic.

There is a fundamental point to be discussed here. At first sight, the problem of depletion and that of global warming seem to be antithetical, and one might argue that depletion would reduce, or even eliminate, the warming problem. If we are running out of fuels, after all, how can we continue producing global warming? The problem has been debated and discussed, and several authors have expressed the opinion that, yes, depletion could prevent a climate catastrophe of the kind envisaged by scenarios based on paleoclimatology.[40] However, the uncertainties involved in these calculations are large and depend on parameters whose value cannot be determined with certainty. Furthermore, most available climate models don't take into account the possibility of the "methane catastrophe," which would greatly accelerate the warming process and make it basically independent of anything that humans could or would do in terms of reducing carbon emissions. Most of these models consider greenhouse gas emissions as an exogenous parameter that is, by definition, not affected by feedback effects generated within the model. A recent review concludes, "The peaking of fossil fuels should not be seen as something that automatically solves the issue of anthropogenic climate change."[41]

What we can say is that, at present, despite the slowdown of the world's economy, we are not seeing a change in the trend of increase in the CO_2 concentration in the atmosphere. What may be happening, instead, is that the increasing need to extract dirty and expensive resources, such as tar sands instead of conventional oil, is reducing the efficiency of the production process and generating more CO_2 for the same amount of energy produced. (See "From Shale Gas to Tar Sands Oil.") What we are seeing is also a worrisome trend of increase in methane emissions from the northern regions of the planet.[42] Even in the long run, it seems unlikely that depletion alone can save us from global warming.

Global warming is just a partial description of what is happening to the Earth's ecosphere as the result of the dispersal of large amounts of minerals resulting from human mining. Even the term "climate change" is only partially true, in the sense that it does not fully include effects such as sea level rise and ocean acidification, as well as reduction in biodiversity. A better term would be "ecosystem disruption" or even "ecosystem destruction." Either term would better convey the combined effects of the avalanche of pollutants that are accumulating not just in the Earth's atmosphere, but also in the geosphere and hydrosphere. It is obvious that for humans the preservation of their subsistence system should be an extremely high priority. If nothing else, the urgency of the situation should be easy to understand by looking at how droughts affected agriculture in 2012, causing great damage to crops and to all cultivations. However, despite the mounting evidence, at present humans do not seem to be poised to reduce their negative impact on the ecosystem by better managing the way they use the planet's mineral resources. The long-term consequences of this inaction remain to be seen and, unfortunately, can hardly be positive for humans and for the ecosystem in general.

The Anthropocene

It is now commonly accepted that human activities have deeply influenced planetary conditions, a fact that has led many to consider the "age of man" an entirely new geological era—the Anthropocene, the era when human activities have had a significant impact on the Earth's ecosystem. The concept of the Anthropocene was introduced and popularized mainly by Paul Crutzen.[43]

Though the Anthopocene is not yet part of the official nomenclature of geological eras, it is now commonly accepted that human activities have deeply influenced planetary conditions, beginning not just with the industrial age, but much earlier. One hypothesis holds that agriculture has deeply affected the Earth's climate since approximately 8,000 years ago.[44] Methane emissions and deforestation from farming may have prevented the Earth from slipping back to a cooling phase that would have resulted in a new ice age, similar to the ones

From Shale Gas to Tar Sands Oil: A Look at the Natural Gas Revolution and the Nonconventional Resource Boom

Ugo Bardi

News about the United States' natural gas revolution has been all the rage in recent years. Industry officials and some economic forecasters have been predicting an abundance that may last decades. Indeed, extracting nonconventional gas from shale formations (for shale gas), coal seams (for coalbed methane), and low-porosity rock (for tight gas) through hydraulic fracturing, better known as "fracking," has revolutionized the fossil fuel market in the United States, increasing production considerably. In the wake of this gas revolution, other nonconventional hydrocarbon resources—including shale oil, tar sands oil, and biofuels—are taking an increasing share of the fuel market in the United States.

This wave of optimism seems to be having a remarkable influence on the public opinion all over the world, creating an atmosphere of enthusiasm not unlike the one surrounding the dot-com bubble in the 1990s. But how long can this trend last? And is it an approach that can be exported outside the United States? The answer is not obvious, for two reasons. One is that the US experience may not be easily replicated in other regions of the world for geological and economic factors. The other, even more important, is that nonconventional resources are nonrenewable and limited, and their abundance may have been substantially overestimated.

The US Story: A Global Production Model?

The rising US production trends are often seen, nowadays, as a refutation of the Hubbert theory that the production of fossil fuels follows a nearly symmetric, bell-shaped curve that peaks and then declines. That peak, when applied to global oil production, is generally called peak oil. Like all models, the Hubbert one is only an approximation, but it is true that the production of fossil fuels and of other mineral resources shows a tendency to follow the model.[45] One of the best-studied cases is that of oil and gas in the United States.

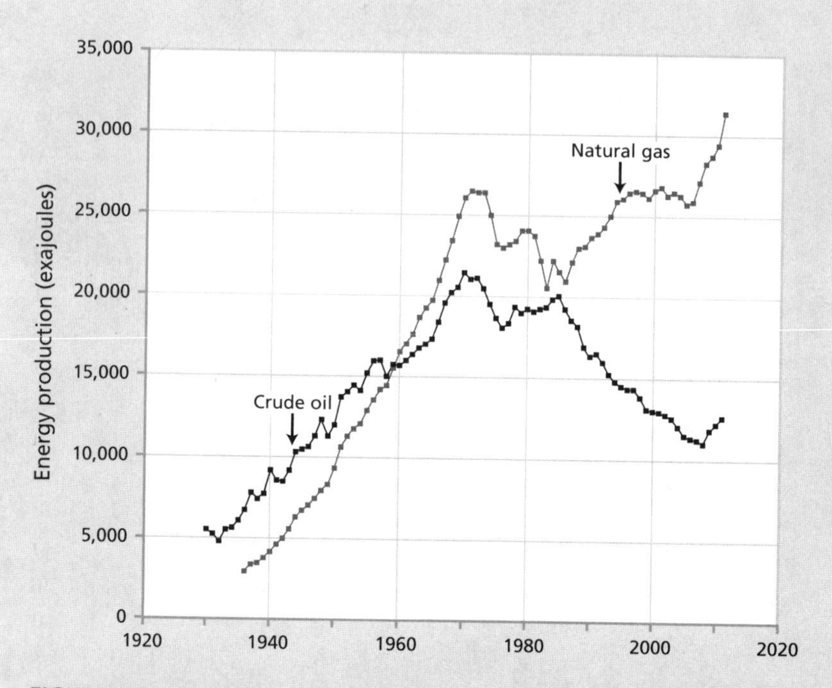

FIGURE 6.6. Production of natural gas and crude oil in the United States (including Alaska). The data for crude oil and natural gas are compared by expressing both in terms of the same energy unit, rather than the customary "barrels" and "cubic feet." The unit used is the exajoule (EJ), equal to 10^{18} joules. One cubic foot of gas is assumed to be equal to 1.1×10^6 J (joules), and one barrel of oil equal to 6.1×10^9 J.

As figure 6.6 shows, oil and gas production in the United States peaked together around 1970. Over the years, oil closely followed the Hubbert curve, with a relatively smooth production decline after the peak—until 2008, when growth restarted due to the exploitation of nonconventional oil. Gas started deviating from the Hubbert curve at around 1985, when production began growing again, until it reached a new maximum at around 2000. The subsequent decline was again interrupted by a sharp upward turn that started around 2006–2007 and that is still ongoing. The Hubbert decline, apparently, is not necessarily forever.

But why did oil production in the United States follow the Hubbert model so much better than gas? To answer this question we must take into account that the model is based on the concept that extraction costs gradually increase because the industry tends to extract the easy-to-access, low-cost resources first.[46] With progressively increasing costs due to depletion, profits shrink, and as a consequence the ability of the industry to invest in new resources shrinks as well. The eventual result is peaking and decline.

A consequence is that after-peak declines can be reversed either by reducing the cost of extraction, for example by improved technology or by increasing investments in extraction. This must be what happened in the United States with gas in the 1980s, and with oil a few years ago. But which of the two factors is the most important? More investments or cost reductions? On this point the general opinion is split between two different narratives. One is that new technologies, the result of human ingenuity, have opened up a new and so far unexplored territory—a vast batch of resources that will lead us to decades, if not centuries, of abundance. The other is that the mechanisms of the economy are trying hard to maintain the production levels of a commodity that is indispensable for our society—liquid fuels—and to do that resources will continue to be removed from other sectors of the economy and used for boosting production for as long as possible.

The majority attribute today's jump in nonconventional resource extraction to the first narrative—new technologies. However, there is no clear evidence of recent technological breakthroughs in this area. Fracking has been around for a long time; it was already in use in the late 19th century, though it was employed with far less intensity than it is today.[47] Directional drilling was developed in the 1930s, and horizontal drilling started to be used in the 1960s.[48] If we look at other nonconventional resources, we see that the exploitation of the Athabasca oil sands in Canada was already under way in 1967.[49] The methods for producing biodiesel were known in the 1930s, while ethanol production has been known since the times of ancient Egypt, at least.[50] Surely there have been cost reductions related to incremental improvements in these technologies, but there is no evident correlation with the discontinuities observed in the production curves.

In economic terms, however, we know that increasing oil and gas prices mean higher profits for the industry, and this may offset the increasing costs of today's more expensive extraction practices. Of course, high prices also tend to depress demand, but energy is such a vital commodity that, within some limits, the industry can maintain high prices and high profits, and keep investing in new, high-cost production. In the end, such market factors can overrule the tendency of production to decline when resources grow scarcer and harder to extract, making the curve deviate from Hubbert's.

But why has US gas production shown such a different trend than US oil production during the past decades? Both are critical commodities for the economy, and in both cases demand is strongly inelastic. But there is a difference: oil can be imported from overseas by tanker at low cost. Gas, instead, requires expensive facilities. As a result, after the peak in 1971 the cheapest route for the US economy to obtain crude oil was to import it, and hence there was no need to pay the high cost of developing new domestic oil resources. Oil consumption did not decrease, but imports grew gradually and today account for almost twice as much as the national production. Only in recent times have increases in oil prices made it profitable to begin extracting large quantities of oil from the Bakken shale formation, in North Dakota and Montana.

The opposite holds for natural gas. Because it is expensive to import, it has made sense to invest in developing new domestic resources, even expensive ones, to maintain the national supply. Two price spikes for gas occurred in the US market: one around 1982–1985, the other around 2006–2008.[51] Both spikes were followed by an increase in gas production—prompting a remarkable increase in the overall number of exploratory rigs in the United States.[52]

So it appears that the present trend of gas production in the United States is no different in economic terms from the ones experienced in the 1980s, even though it exploits geologically different resources. If the market is willing to pay a high enough price for a mineral commodity, then a technology able to produce it will be deployed—even if that technology has been known for a long time, having formerly been considered too expensive to use.[53]

The downside of the present trends is that in the United States gas and oil are being extracted at rates that have never been experienced before, and that means the national reserves are being rapidly consumed. In fact, recently the number of gas rigs has been falling rapidly.[54] Because it is necessary to drill with those rigs before they can produce gas, this fall signals a coming decline in gas production. Indeed, it appears that the decline in gas production in the United States is already starting.[55] Gas prices have also been going down during the past few years, and the still-rising production and investments are looking more and more like a speculative financial bubble ready to burst in the near future.[56] A similar trend is observed for the production of shale oil in the United States, mainly as the result of exploiting the Bakken shale formation. Also in this case, there are elements indicating that production may be close to reaching a plateau and subsequent decline, although this may be farther ahead in the future than it is for natural gas.[57]

If we look at other regions of the world, the estimates of how much nonconventional gas might exist are very uncertain. In Europe these resources are most likely modest. China, on the other hand, may have resources comparable to those of the United States.[58] Both Europe and China are much more densely populated than the United States, though, and the ecosystem impact of procedures like fracking may be so large that it would be either extremely difficult or impossible to extract nonconventional gas because of political opposition and legal requirements.

In principle natural gas can be transported overseas by cryogenic liquefaction, and regions like Europe could profit from the shale gas boom in the United States even without developing resources of their own. However, the US productive phase appears to be too short-lived to justify the time and the investment needed to build the expensive transport infrastructure that would be needed. Hence, importers such as western Europe are likely to remain linked to traditional gas resources and to the complex network of pipelines that transport them. Because geopolitical factors weigh heavily on the management of these pipelines, the future gas supply for importers, and in particular for western Europe, remains fraught with uncertainties. There won't be a worldwide peak for gas, as there has been for crude oil, but depletion will manifest itself as local crises, which the public will most likely interpret as politically driven. Nevertheless, in the long run there is no escaping the fact that we are dealing with a nonrenewable, limited resource.

The Ecological Impacts

Shale gas is often described as a "bridge resource" that should help us transition from fossil fuels to renewables or, in some views, a combination of renewables and nuclear.[59] Indeed, for the same energy produced, natural gas generates lower emissions than coal because of its different chemical composition and because it can be used to fuel gas turbines, which are often more efficient than the steam turbines normally used in coal-fired plants.

However, a true comparison considers not just the energy content of the two different fuels (coal and gas) but also the energy needed to produce them and the losses incurred in the process. It has been claimed that shale gas may be more damaging in terms of greenhouse gas production than coal because of the losses involved during extraction and the fact that methane, the main component of natural gas, has a stronger effect than carbon dioxide as a greenhouse gas.[60] This interpretation has been debated and criticized but cannot be ignored.[61] We must also consider the fact that a higher cost of extraction implies that more resources—including energy—must be expended. In the end, this lowers the efficiency of the exploitation cycle and generates more pollution per unit of energy produced, regardless of how efficient the fuel extract is in generating energy in a power plant. This reality implies that nonconventional gas, just like natural gas in general, cannot be considered as a solution for the present climate crisis. In fact, the trend toward lower-yield fuels is going to *increase* the human impact on climate, in part by diverting resources that could be used to develop and deploy more climate-friendly energy sources. And as we increase gas and oil production, we also increase the depletion rate of these limited resources.

Hence, the "new age" of fossil fuels will most likely turn out to be a short-lived bump on the path toward unavoidable decline.

that have prevailed during the past million years or so. This hypothesis cannot be considered proven, but it is an indication of the important effect that human activities have on climate.

In any case, the effect of ancient agriculture on climate pales in comparison to the extent of the modern industrial and agricultural activity. The worst effects of the great mining experiment performed by humans are still to be seen. We can't know the details of what is going to happen, but we do know that human mining has transformed—and is still transforming—the Earth into a different planet. It is a planet whose surface and ecosphere have changed as the result of mining. We are heading, it seems, toward the climax of the Anthropocene, a new era in which the atmosphere contains greenhouse gases in amounts that haven't been seen in millions of years. Toward a planet where the oceans are acid and the polar ice caps have shrunk or disappeared altogether, where temperatures have risen to levels so high as to exterminate most vertebrates in equatorial and tropical regions, where acidification has destroyed most marine life, and where ocean rise has swamped most human-made coastal settlements. Whether we will be able to live on this new planet is impossible to say.[62] It may be that a dark era is awaiting us: the dark side of mining.

PART THREE

A NEW PLANET

The growth of photovoltaics and wind energy has been impressive over the past two to three decades, with the two fields doubling their output about every 5 to 10 years. At these rates both wind and photovoltaics could reach the goal of 1 terawatt each installed by around 2020. Does that mean we can substitute for fossil fuels with renewable energy? It depends on what we mean exactly by "substitution."

7

The Red Queen's Race:
The Future of Civilization

The Western Roman Empire waned for at least three centuries before it disappeared in the 5th century. During this long decline, emperors must surely have understood that something was wrong, but their only response was to struggle to maintain the status quo. In a way, they were running the Red Queen's race, the useless struggle described in Lewis Carroll's *Through the Looking-Glass*. Everyone in the Red Queen's kingdom had to run as fast as they possibly could just to stay in one place. As the bewildered Alice noted, it was a lot of work to get nowhere.

The Romans, too, were making a tremendous effort to stay exactly where they were, avoiding all changes in the political and economic structure of the empire. But for them it was a lost race. The natural resources that had sustained the imperial system were running out. The Spanish mines had ceased to produce gold and silver. There were no more neighbors that were easy to overrun and pillage. Agriculture was suffering as fertile soil eroded away. The enormous expenses for the army, for heavy fortifications, for the imperial court, and for a huge bureaucratic system required a taxation regime that, eventually, bankrupted the Roman society. The same bureaucratic structures that had ensured the empire's growth had become a burden, yielding diminishing returns and an inability to cope with decreasing resources.[1]

Yet the solution to the empire's problem was staring right at them: the Middle Ages, which followed on the heels of the empire's collapse. The new era brought new ways that freed Europe from the suffocating imperial bureaucracy, from the enormous expenses to keep armies and fortifications, and from the terrible tax burden that was destroying the very fabric of society. During the Middle Ages defense was provided by local militias and taxes were paid only to local rulers. After a period of economic contraction that allowed soil to replenish and forests to regrow, Europe could rebuild its agricultural prosperity and restart its cultural and economic growth—a necessary precursor to the Renaissance.

Roman emperors never understood the need for sweeping changes in the way the empire was run. We can say that they fell victim to the human tendency to "pull the levers in the wrong direction," a concept that Jay Forrester developed and Donella Meadows built upon in her now classic reference for system change, "Leverage Points: Places to Intervene in a System."[2] As Meadows put it, "These are places within a complex system (a corporation, an economy, a living body, a city, an ecosystem) where a small shift in one thing can produce big changes in everything." The problem is that while most people clearly understand what the leverage points of the system are, they tend to miscalculate the final result of the feedback cascade and pull the system levers in the wrong direction. So the Roman emperors kept pulling the levers in the wrong direction: they enlarged the army while they should have reduced its size, they created more bureaucracy when they should have created less, and so on.

The plight of the ancient Roman Empire looks much like our present situation. We also have problems of diminishing natural resources, excess bureaucracy, and all sorts of pollution problems. That doesn't mean that we have to expect that the solution for our problems will be new Middle Ages, complete with armored knights and fortified castles. It does mean, however, that we have been running the Red Queen's race, too. We have been pulling the levers in the wrong direction. We are expending great effort to maintain things as they are, without realizing that the only way out of our predicament is to embrace change instead of fighting it.

Many times our awkward attempts to solve problems are in reality just ways to postpone the need to face them. And postponing a problem can lead to a great tangle of unintended consequences down the road. That tangle can grow ever more complex when a society is facing not a single problem but rather several interacting ones—as now, when we are confronting climate change, peak oil, growing human population, and mineral depletion. What will happen, for instance, if we react to oil depletion by drilling deeper, drilling more, drilling in places where no one has drilled before? The "drill, baby, drill" mantra made famous by former Republican vice-presidential candidate Sarah Palin gets cheers from those focused on short-term gains. But, obviously, drilling more leads to faster depletion; it is a classic case of pulling the levers in the wrong direction. At the same time, drilling more worsens the climate problem because it takes lots of greenhouse-gas-emitting fossil fuels to reach those hard-to-access reserves, and once the new oil is unearthed and burned, even more greenhouse gases are generated.

So the first thing we need to clarify is what we mean exactly when we talk about "solutions" to our present problems. If we mean that we want to use a mix of better technologies and virtuous market forces to keep everything as it stands now, including SUVs and vacations to Hawai'i by plane, we need to pause and reconsider. That approach is simply not workable. Technology can do a lot of things, but it cannot change the laws of physics. We cannot win the Red Queen's race, and there is no escaping the fact that we will see great changes. Whether we accept these changes or fight them is, in the long run, immaterial because change is unavoidable. What we can do is to examine how these changes might occur and what kind of world we may be moving toward.

Mineral depletion plays a leading role in that future world. So, if we need to reduce our dependence on conventional mining, reduce the impacts of mining on the ecosystem, and at the same time maintain some kind of a working industrial economy, we can think of three possible approaches:

- Substituting common minerals for rare minerals
- Recycling and reusing minerals
- Reducing the consumption of all mineral commodities

The Substitution Option

Substitution is popular as a remedy for depletion. The idea is that, as we run out of a rare mineral resource, we can always substitute another, more abundant one. Sometimes rising prices of one resource will jump-start the production of a substitute to replace it—as Hotelling proposed with his "backstop resource" concept.[3] Those who follow this line of reasoning also normally propose that, soon, new technologies will bring down the cost of producing the new resource, making it more and more available. In this way, resources appear to grow as we consume them. If this were true, we could substitute our way to winning the Red Queen's race. And if we can substitute everything, why should we be worried about depletion?

Substituting is not so easy as some economists would lead us to believe, but the concept is valid within physically reasonable limits. Perhaps the first study that examined this issue was conducted in the 1970s by H. E. Goeller and Alvin M. Weinberg, resulting in their "principle of infinite substitutability."[4] A name like that makes the concept seem outlandish, but their conclusions were based on sound physical principles and valid within the limits of their initial assumptions. "Infinite," in this context, doesn't have to be taken in a literal sense; consider it to mean that a combination of technological inventiveness and the growing availability of energy would permit humankind to overcome the depletion problem for a very long time to come.

Goeller and Weinberg started with the idea that gradual depletion would lead to the impossibility of mining rare mineral elements at reasonable costs. They proposed substituting these rare elements with ones that are common in the crust and that can be extracted using reasonable amounts of energy that, they assumed, would come from a new generation of nuclear power plants.

The general validity of the substitutability option was based on examples like copper—a rare element in the Earth's crust and one that we use as the main electrical conductor in all sorts of devices. Electrical-conducting materials are indispensable in our society, so can we substitute copper with something more common? Yes, we can use aluminum, which is almost as conductive as copper and quite common in the crust. Yet aluminum's chemical and mechanical characteristics are much different from those of copper. It may be a good conductor, but it is brittle, it oxidizes easily, and, more importantly, when made into thin wires it can catch fire if heated. So, if you are rewiring an electrical system, for instance, aluminum wiring can't easily substitute for copper wiring. Indeed, wiring homes with aluminum had become popular in the United States in the 1970s, but most of these homes were later rewired

FIGURE 7.1. A picture of the slurry tanks of the aluminum plant of Butzflethermoor, in Lower Saxony, Germany. This image provides some idea of the environmental impact and cost of these plants. Substituting copper with aluminum is possible, but not painless.

with copper because of short circuits, overheating, and fire.[5] Finally, aluminum production is energy-hungry. According to the available data, producing aluminum takes more than four times the energy needed for the same amount of copper (by weight).[6] And because it doesn't conduct electricity as well as copper, even though aluminum is a less dense metal, it still takes about two times more energy to manufacture an aluminum wire able to carry the same current as a copper one.

The problems with the copper/aluminum case are typical of substitution in general: it is often possible but never straightforward, and often it is energy expensive. We find these problems in other examples that Goeller and Weinberg discussed. For instance, stainless steel is an alloy of iron and (mainly) chromium. Iron is abundant in the Earth's crust, but that's not the case for chromium. If chromium depletion forces us to look for alternatives, we could use titanium as a substitute for stainless steel. Titanium is abundant and suitable for structural applications where a good resistance to corrosion is necessary. Unfortunately, though, titanium has a high melting point, requiring large amounts of energy for its production. Again, the substitution strategy turns out to be power hungry.

When Goeller and Weinberg conceived their principle of substitutability, they were banking on an ample future supply of nuclear energy. However, things are much changed from those times, and it now seems clear that the wide expansion of nuclear power they envisaged is not going to materialize in the near future, and perhaps never. So, if we want to be able to cope with depletion via substitution, we have to revise our thinking. We can certainly make substitutions if enough energy is available. The problem is that most of the energy we use today comes from fossil fuels, and our supply of fossil fuels is decreasing. What do you substitute for fossil fuels? There's not an easy answer. For instance, hopes of replacing fossil fuels with biofuels are based on a gross misunderstanding of the efficiency of photosynthesis and of the needs of agriculture.[7] However, the latest generation of renewable energy technologies, such as solar and wind power, look more promising. While some of these technologies still have economic and adaptation hurdles to overcome, the growth trends of renewables such as photovoltaics and wind have been impressive, currently doubling at a rate of about once every 5 to 10 years—a pace never seen before for any energy technology.[8] Of course, nothing can grow exponentially forever, and the most recent data appear to indicate a slowdown in the trend. Nevertheless, we are truly seeing an energy revolution: renewable power has a market and it grows. But can we use renewables to substitute for fossil fuels?

The feasibility of a large-scale diffusion of renewables is currently the object of a heated debate, with favorable opinions as well as more cautious ones.[9] Basically, the large-scale diffusion of renewable energy depends mainly on two factors: their energy efficiency (expressed in terms of EROEI, energy returned for energy invested) and their conversion efficiency, expressed in terms of the land area needed to obtain a given amount of energy. It is likely that the EROEI of the most rapidly growing forms of renewable energy (wind and photovoltaic) is still lower than that of fossil fuels.[10] However, renewable-energy technology is rapidly evolving: the EROEI of some advanced thin-film photovoltaic technologies may already be larger than that of present-day fossil fuels, and technology and scale are causing the EROEI of wind energy to rise as well.[11] At the same time, the EROEI of fossil fuels keeps diminishing as depletion forces the industry to extract from more and more expensive deposits.

The question of conversion efficiency, then, is crucial in determining how much energy renewables can produce without negatively impacting agriculture or other uses of the land. Here, modern solar and wind technology have become efficient enough that we can calculate that it is theoretically possible to deploy a mix of renewable technologies over the land in such a way as to produce as much energy as is produced nowadays by fossil fuels without major negative impacts on agriculture.[12] Such an infrastructure could actually be accommodated mainly on nonproductive or already built/paved land.[13]

All that doesn't mean that renewable energy can substitute fossil fuels in all their tasks, such as, say, providing fuel for SUVs, nor that the energy produced can have the same low cost that the golden age of fossil fuels has made us accustomed to.[14] What it means is that renewable energy can insure that, at least, we won't be forced go back to a purely agrarian society as the result of fossil fuel depletion. It also means that we'll still have some energy that will permit us to continue mining within some limits, even though, of course, a mineral industry based on renewable energy will have to be rethought and reorganized.[15]

At this point, however, we face a fundamental question: Don't renewables depend on nonrenewable rare minerals themselves? In part, the answer is yes, but it is also true that this dependency may not be crucial, especially if new renewable technologies will be developed with the specific aim of using little or no rare minerals. For instance, photovoltaic cells can be produced using mainly materials abundant in the Earth's crust: silicon, phosphorus, boron, and nitrogen for the cell itself and aluminum for the contacts. The only rare metal used in the present silicon-based cell is silver for the electrical contacts, but that can be substituted with other materials with only a minimal loss of performance. Solar-concentration plants, an alternative to photovoltaic plants, capture solar energy altogether differently: they use mirrors to generate high temperatures in a fluid that is then used to operate a thermal engine. These plants require mainly aluminum or steel for the mirrors and the piping, and steel for the steam turbines that generate mechanical energy.

The mineral-use outlook is reasonable on the wind frontier, as well. Wind towers are mostly manufactured using abundant materials: concrete, steel, and aluminum. However, wind plants rely on magnets to transform mechanical power into electrical power, and today those magnets are made with rare earths. This need poses an important depletion problem, but it is possible to recycle rare earths and also to make magnets without them, albeit with a loss of performance. Generally speaking, all energy-producing plants need electronic systems to control and convert the energy produced into forms that can be easily exploited by users. These electronic systems, as they are now produced, make significant use of rare metals, but it is possible to reduce their use and recycle them efficiently.[16]

One problem with most renewable technologies is the intermittent and seasonally oscillating production—an issue for a population that is used to energy available on demand. That's not an unsolvable problem; storage technologies exist, but they are expensive and require adaptation to our systems. We can maintain baseline production using a combination of non-intermittent renewable technologies such as geothermal, hydroelectric, and biomass energy—all without a critical dependency on rare mineral commodities. We can then manage demand through a smart grid, which would allow users to tap the production of a large number of plants and adapt their demand to the available supply. We are not used to this kind of market for energy, but it is the norm for many areas of commerce. You wouldn't expect, for instance, to just appear at the airport and purchase an international plane ticket "on demand," unless you are willing to pay a lot of money for it. The energy market, too, is trending toward different costs for different times and days. Energy will always be available, but at some moments it will be much more expensive and people will be discouraged from using it. The opposite will occur for moments of high availability and low prices.

Even with energy available, some of the minerals we depend upon will be difficult or expensive to substitute, as we've seen in the glimpses that run throughout this book. For instance, at present there are no substitutes for the combination of noble metals used for automotive catalytic converters. That doesn't mean that we can't find other ways to reduce the pollution produced by internal combustion engines, but to do so we will have to rethink the problem from scratch. Or we could switch to battery-powered electric motors—but as we know, they require lithium, which is also facing a depletion problem. (See "Platinum Group Metals," page 117, and "Lithium," page 128.)

And how about phosphorus? Part of the biological machinery of living beings, it is indispensable in the fertilizer used in conventional modern agriculture. But there is no substitute for it, and the solution to phosphorus depletion lies not in replacing the mineral but in implementing new agricultural methods that can capture the phosphorus that travels through plants and the animals that eat them and then return them to the Earth. (See "Phosphorus," page 163.)

So substitution is often, but not always, possible. Indeed, if substituting some rare mineral commodity were convenient and easy, we would have already done it long ago. Nevertheless, substitution is a good strategy provided that we accept three realities: that we need to invest serious resources in developing substitutes; that not everything can be substituted; and that the world that will result from this process will be fundamentally different from the one we are accustomed to.

The Recycling/Reuse Option

If we could recycle 100 percent of our waste, we'd have no resource depletion problem, and we could go on forever using what we have laboriously extracted from mines in the past. It would also greatly reduce the environmental impact of mining, particularly since it wouldn't put any new materials into the ecosystem. Again, it would be a way to win the Red Queen's race, remaining exactly where we are.

Recycling anything at 100 percent efficiency is not impossible, in itself. After all, plants have been recycling the minerals they use for hundreds of millions of years and, on the whole, never running out of anything. But considering the way our global industrial system is structured, getting even close to 100 percent recycling would require amounts of energy that we don't have today and are unlikely to have in the future.

Of course, the fact that something is difficult doesn't mean that it should not be attempted, and that is, indeed, a goal of many pioneering communities around the world. These communities are trying to manage household waste by instituting zero-waste programs that encourage people to recycle more, reuse more, compost kitchen scraps, and generally avoid products that can't be recycled or reused.[17] These communities—through various networks of municipal agencies and private businesses—collect and separate recyclables, reclaiming impressive amounts; cart kitchen waste to composting facilities that later truck it to farmland; coordinate with, or even operate, repair and resale centers for reusable items; and implement regulations that hold manufacturers accountable for unnecessary waste. Forward-thinking companies, too, are engaged in eliminating waste from their manufacturing processes and packaging, and taking responsibility for end-of-life reclamation for their products.

But these efforts remain limited in extent, while in most places around the world there has never been an effort to manage waste in such a way to make it easy to recover useful materials from it. The goal has been to simply make it disappear from view. Until recently, "waste management" operations simply collected all the mixed-together waste accumulated in bins and threw it into large holes in the ground that go under the name of "landfills." While some form of recycling is generally common now, it is also often minimal and inefficient. So we still tip vast amounts of waste into landfills or burn it in incinerators. We have even, at times, dumped it into the sea—a practice fortunately forbidden by international treaties today.[18] As a consequence, many minerals that have entered the world's economy during the past few centuries are today accumulated inside landfills or dispersed in the ecosystem.

It is a fundamental rule of chemistry that nothing is created and nothing is destroyed. The minerals that we have laboriously extracted from mines have not been destroyed. They are still here, somewhere. But reclaiming the minerals from waste that we have recklessly dispersed around or even dumped into the ocean is a monumental task, and it is unlikely that we could reasonably recover much of them. The minerals from waste that we have sent up in smoke are, for reuse purposes, gone forever. They have a new role as air, water, and land pollution. The solid ash that settles at the bottom of incinerator smokestacks does contain considerable amounts of useful metals, and there have been several attempts to study methods of recovery.[19] But the chemical processes necessary to separate the various components are complex and energy intensive. So far the minerals that can be extracted—such as sodium, potassium, and calcium—are far from being crucial resources for industry. Extracting rare metals from these ashes has proven costly enough to be impossible. Once again we encounter the essential problem of the universal mining machine: it is a question not of quantity, but of energy cost.

Which leads us back to landfills. What if we could recover from landfills what we threw away in them? There are landfills created during Roman times that still smell bad when archaeologists dig them out. Indeed, archeology often consists of the study of what peoples of ancient civilizations discarded. Researchers even study our own civilization by looking at what we discarded in landfills during the past decades or centuries (this is called, sometimes, "garbology"[20]). A modern landfill, if built with adequate precautions, provides long-term waste storage. The question is, though, is landfill mining a practical possibility for the real world?[21]

Many interesting artifacts can be recovered by mining landfills, but when it comes to producing large amounts of basic commodities, the task is expensive, difficult, and even dangerous. The problem is that waste is normally stored in landfills without any thought about future recovery. Not that a landfill is a completely random pile. Modern ones are built in a layer-by-layer fashion to reduce putrefaction and other unwanted effects of waste decay. Since they produce methane as the result of the gradual degradation of organic material in their oxygen-free environments, many landfills are equipped with systems that recover and burn some of the methane to produce electric power. But when it is a question of recovering metals and other minerals, the problem is very difficult. The average concentration of rare metals in a landfill is low and an even more difficult problem is that all metals are mixed together. So exploiting a landfill as if it were a mine would require sophisticated and expensive separation techniques and much energy, and would also create a lot of pollution. However, a landfill is unlike a conventional mine in the sense that metals can be often found in the form of macroscopic objects like aluminum cans, electric wires, steel parts, and so on. That permits methods of separation that wouldn't be possible in conventional mining. For instance, iron objects can be separated using magnetic fields, and aluminum cans can be singled out on vibrating platforms that separate materials of different density. Still, the work poses many safety and comfort problems for workers. Organic waste generates foul odors, and its high bacterial content can cause health problems. At the same time, trash may contain sharp-edged objects, poisons, explosives, noxious gases, and many more potential hazards.

In practice, landfill mining in industrialized countries is almost never economically convenient, except in those cases where the land recovered from an old landfill is extremely valuable, as for instance in urban development projects. Even in these cases, though, it is doubtful that landfill mining makes sense in economic terms in industrialized societies.

The situation is somewhat different in poorer societies, where recovering waste is a traditional job for the destitute. These workers—sometimes officially recognized but mostly not—are in some ways like the gleaners of centuries past who optimized agricultural yields by collecting crops left in fields after harvest. It is an ancient way of resource management. It also requires little expense of resources and energy.

FIGURE 7.2. Informal recycling workers in southern nations perform a useful task, but they work in difficult and unhealthy conditions and the importance of what they are doing is often not recognized.

This participatory form of waste collection is showing signs of spreading around the world.[22] However, it encounters resistance from several sectors. The established waste collection agencies see efforts to organize workers, formalize their roles, or assist in their efforts as competition for their own services. Others contend that relying on the poor to recycle waste keeps them in long-lasting poverty. Governments tend to view with suspicion and usually forbid anything that they are not sure to be able to control. So, in almost all cases, *catadores, cartoneros*, and *binners*—as they are called in various countries—are usually poorly paid and their activities are often expressly forbidden by law.[23]

Especially when waste pickers are not protected by laws and formal organization, their job is dangerous, dirty, and heavy and often involves child labor. Nevertheless, those who have recognized the value of waste pickers to society advocate for better conditions. First and foremost, advocates say that their role must be recognized and suitable legislation must protect the workers from the hazards of their job, as described in "Replenishing the Earth through Informal and Cooperative Recycling." If that happens, we may see a small revolution in waste management that leads to recovering more of what we carelessly throw

Replenishing the Earth through Informal and Cooperative Recycling

Jutta Gutberlet

Increased mass consumption, planned obsolescence, and throw-away attitudes have expanded the scramble for mineral resources worldwide. And the more we consume, the more waste we generate: throughout the production process, in packaging, and when we finally discard used products. Most of this waste is not recovered for reuse or recycling, thanks to an economic development model that has produced a society based on a perception of unlimited growth and unrestricted natural resource availability.

However, a growing number of theorists are proposing radical changes in how we pursue economic activity, emphasizing social and environmental justice, an end to endless growth, and a transition to sustainable community development.[24] From this standpoint, recovering resources for reuse and recycling becomes essential. Not only can it lessen the burden on limited mineral resources by recovering raw materials versus mining them, it can also help transform economies—generating new enterprises to recover and reuse items from the waste stream, putting more people to work, linking local authorities, private enterprises, the state, and citizens in zero-waste efforts, and redirecting us toward ethical and sustainable consumption.[25]

The prevailing disconnect between products and their embedded resource value, energy, and labor input generates a purely utilitarian relationship between consumer and product. Once a product's life is considered over, all too often it gets thrown away. Current waste management options, and particularly waste incineration (more recently euphemistically termed "waste to energy"), facilitate this behavior by taking away from both the consumer and the producer the responsibility for the environmental impacts caused by waste.

It will take much education and awareness building to get people around the world to understand that there is no such thing as waste and that every cause has an effect and every action is accompanied by a reaction. Consumption drives our current economy. But it will take significant changes in lifestyles and habits, emphasizing sufficiency and simplicity over abundance and consumption, to make the necessary change to a low carbon-energy system and to curb our ravenous extraction rates.

So, while recovering the materials embedded in solid waste (recycling) is critical, most materials are not endlessly recyclable, and recycling processes also require energy, water, and often the input of additional resources. Recycling also creates burdens through collection, reclamation, and transportation. Hence, we must follow the strategies of avoidance and reuse, not just of recycling.

Recovery Efforts Under Way

Nowadays landfills can be considered urban mines. Decades of material disposal have accumulated in the waste flow, including metal-containing products. A global stockpile of 225 million metric tons of copper, for instance, is thought to reside in landfills.[26] Copper and other metals—particularly iron, steel, and aluminum—are regarded as being easy to recycle and have a long tradition of being partially recovered in the industrial cycle. Iron and steel have an end-of-life recycling rate between 70 and 90 percent.[27] Other metals have much lower end-of-life recovery rates; manganese, niobium, nickel, and chromium are at 50 percent.[28] Several of the nonferrous metals, including lead, aluminum, and copper, are reported to have a recovery rate of over 50 percent, and estimates for magnesium recycling range between 25 and 50 percent.[29] The importance of recycling activities is illustrated by the largest municipal recycling facility in China, which recovers twice as much copper per year as the annual production rate of the largest copper mine in China.[30]

Precious metals such as gold, silver, and platinum are widely recycled, given their high value. However, some metals are hardly recycled. Many high-tech products, including computers, mobile phones, solar panels, catalysts, and batteries, employ low concentrations of specialty metals like gallium, indium, and rare earth elements. The end-of-life recycling rates for most of these specialty metals lie between zero and 1 percent.

Overall recycling rates, including the recovery of metals and minerals, are increasing worldwide as prices for raw materials and costs for disposal rise and as environmental pressures increase. In general this has been good news not just for the environment but also for communities and local economies.

Around the world, local governments, private entrepreneurs, and other organizations are building new networks for waste recovery, reuse, and resale operations, many of them aimed ultimately at creating zero-waste streams that recover and reuse everything from product packaging to kitchen scraps. A huge part of that effort depends on getting manufacturers to design products with less excess and more recoverability in mind.

In many parts of the world, the recovery of discarded minerals also often depends on informal and cooperative recycling. In the Global South informal recycling recovers large proportions of resources, employs significant numbers (approximately 1 percent of the population), and generates livelihoods for the poor. Almost one million people are involved in resource recovery in Brazil, for instance. A small proportion of these recyclers, called *catadores*, are organized in associations and cooperatives. Yet most of them still work under deplorable conditions. Some municipalities actively involve *catadores* in the collection and separation of recyclable materials by granting space to set up triage centers, providing transportation, supporting them in capacity development, or paying them to collect and redirect what might otherwise end up at the landfill.

There are many different ways *catadores* can collect waste, ranging from scavenging at landfills, open dumps, and garbage placed at the curbside to recovering materials directly from households, industries, or offices or organized door-to-door collection of source-separated materials, as part of the city's waste management program. Itinerant buyers may purchase source-separated recyclables such as bottles, cans, paper, and cardboard from residents. Fixed buyers, essentially middlemen, might also buy the material from informal recyclers. Finally, a large array of different industries and transportation businesses, small to large in scale, are involved in the recycling activities.

The work of the *catadores* is commonly diurnal. Materials are collected, sorted by category (plastic type, glass color, paper or cardboard type, and metals), and sold to intermediaries. When large quantities of separated materials are involved, they sell directly to industry. In Lima, Peru, for example, the municipal government recovers merely

0.3 percent of the city waste, but that number jumps to 20 percent with the informal collectors involved.[31] Likewise, 37 percent of the population in Santa Cruz, Bolivia, is serviced by informal waste collectors only.[32] In Delhi, India, only 34 percent of the city's refuse is recycled, and 27 percent of that recycling is performed by informal collection services.[33]

The work of recyclers, whether informal or organized, is often underappreciated. There is no doubt, though, that without them many resources would be lost and cities would have to deal with even more garbage, on a daily basis. Informal recyclers perform such a substantial favor to the economy that cities such as Bangkok, Jakarta, Kanpur, Karachi, and Manila save upwards of $23 million annually through the recyclers' work.[34] These savings come in the form of reduced import costs of mineral commodities and savings on waste management systems.[35]

Salvaging materials considered to be "waste" also prevents further strain on the environment, and not just in terms of mining. In energy-intensive industries, the recovery of basic materials such as aluminum, steel, paper, and iron can result in a huge amount of energy savings. Aluminum recycling can save up to 95 percent of the energy costs required in the production of virgin materials. Steel recycling can yield a 40 to 75 percent savings in the amount of energy required to produce steel.[36] The recycling of packaging containing metal products and the re-smelting of used metal packages, such as beverage cans, also contribute to these environmental gains.

So a key to continued success lies in supporting *catadores* and other informal recycling workers. In Latin America, and more specifically in Brazil, recycling cooperatives have been established to generate organized work opportunities, which allow for peer support and provide collective learning opportunities. Cooperatives are also more likely to offer regular pay.

Brazil has passed federal laws recognizing the profession and allowing recycling cooperatives. But while the formal assistance to recyclers has proven beneficial, it has also created a level of dependence on political parties. This dependence is problematic, because a change in government can lead to a number of challenges—such as when assistance to the cooperatives is retracted, or the goverment pressures them to streamline into more business-oriented forms of work. In some cases cooperatives have been weakened as a result of governments giving priority to other forms of waste management. In several Brazilian towns, for example, the work of the recyclers has been severely impacted by the official decision to set up incineration ("waste to energy") plants, which burn waste (destroying the raw resources that went into making the discarded items) rather than recovering and recycling it.[37]

On the positive side, some municipalities have strengthened their commitment to inclusive solid waste management, officially paying for the recyclers' service and providing infrastructure support. And some cooperatives have recently started to recover plastics and metals from electronic products, as well as platinum group metals. (However, capacity-building programs are needed to prevent occupational health problems for those working with hazardous materials.)

The Road to Zero Waste

Whether organized, unofficial, or cooperative, resource recovery efforts can offer huge potential for social and environmental gains. Enabling the transition away from wastefulness will require investments in organized selective waste collection and separation, reuse, and recycling, as well as environmental education about zero-waste goals. And the responsibility for driving change rests not only with governments and businesses, but also with citizens, who can choose to consume more wisely and recycle more fully.

away. Nevertheless, even with these improved methods, recovery from landfills cannot alone recycle more than a small fraction of the waste produced by the industrial society.

The problems involved with the "downstream" recovery of mineral resources from industrial cycles has led to efforts to improve the "upstream" management—that is, to create waste that is not simply a random mixture of everything. In urban areas of industrialized regions of the world, people are usually asked to separate their household waste into different streams, such as paper, plastic, metal, organic, and glass. However, the various streams that arrive from separated collection to the processing plants are often not pure enough to produce new feedstock at market prices. That doesn't mean that these efforts should not be performed; a poor separation of waste is always better than no separation at all. But the difficulties and the costs involved are large, and the separated collection of urban waste alone is also not going to solve the mineral depletion problem.

A basic problem with waste recycling is "downcycling." This term refers to the fact that the recycled material is normally of lower quality than the same material manufactured from pristine mineral sources. Let's again look to steel for an example. All over the world, there exists a well-established recycling system for steel. Today 68 percent of all iron and steel produced is recycled.[38] That's surely a good thing, but there is a problem: most of the steel produced today is an alloy of several different metals, determined by the desired application. Steel may contain chromium, cobalt, silicon, manganese, vanadium, and other elements, depending on how hard, strong, or corrosion-resistant it needs to be. Recycling steel means fusing together objects, such as old cars, scrap metal, household appliances, and the like, which contain different kinds of steel. Controlling the concentration of the extra metals during processing is so complex and expensive that it is impossible in most cases. As a consequence, the composition of recycled steel is normally an average of the compositions of the different input steels. Such an alloy is a lower-quality product suitable only for nondemanding applications.

There are many more examples of downcycling. Each time paper is recycled its fibers get shorter, resulting in an inferior product. When many different kinds of plastic get mixed together, they provide a product of poor mechanical properties that has very limited uses. Aluminum is mixed with magnesium in beverage cans, and the recycled product requires additional stages of separation if it has to be transformed into pure aluminum. Glass recycling is easier, but there are problems also with the variable amounts of types of oxides that are contained in commercial glass products. Harvesting recyclable minerals from electronic waste has its own challenges, but it has the advantage that, at least, electronic devices are easily recognizable and recoverable in a compact form. Some metals can be recovered in a nearly pure form from electronic devices, such as gold and silver from printed boards. Unfortunately, that's not always the case, and economic factors, at present, prevent the recovery of less valuable materials. For instance, silicon could be theoretically recovered from printed circuit boards, but at present, it is not. If it were, silicon would also suffer from the downcycling problem, being "doped" with different elements for different applications.

In all cases, recycling becomes more and more difficult as the recycled fraction increases and as higher performance is required in the recycled material.

This greater difficulty translates into greater financial and energy costs, and as a result the amount of recycling that turns out to be economically convenient is much less than 100 percent. According to the US Geological Survey, the average recycling rate for most metals in the United States is about 50 percent.[39] Lead has the highest recycling rate, at 74 percent. Iron and steel follow at 60 percent. Other common metals are recycled at lower levels; copper and aluminum don't do better than around 30 percent. Rarer metals are recycled at lower rates, and some, such as indium and gallium, are not recycled at all. Clearly these recycling rates are too low to solve the depletion problem. Even if we could perform multiple cycles (and normally we can't, because of the downcycling factor), imagine the case of a metal that is recycled at 50 percent. After just four cycles, we have already lost almost 95 percent of the original amount!

We can only conclude that recycling may have a significant effect in fighting depletion only if we change the way industrial production is carried out, so that it produces less waste, designs products that are easily dismantled and therefore ease the recovery of the materials they contain, and yields products with a long service life. Restructuring the industrial system in this way is sometimes called "closing the industrial cycle," or "the circular economy," or "cradle-to-cradle" (C2C) design.[40] The principles involved in these methods, and particularly in C2C design, include the concept that "waste is food," meaning that *all,* not just a fraction, of the materials used in industry can be reused. In an open system, subjected to an external flux of energy, it is perfectly possible to "close the cycle" and recover everything. All these ideas are making inroads in society, though at present only in niche markets.

Today, it is rare to find legislation that requires companies to recover the materials used in the products they manufacture—with some exceptions, such as for some electronic waste, as "Electronic Waste and Rare Earths" describes. But industrial design still mostly relies on the concept of planned obsolescence, which involves designing objects with the specific purpose of making them impossible to repair, thereby forcing customers to discard them and buy new ones. In addition, there is a widespread opinion that market forces will solve all problems once they appear, making legislation unnecessary. But the market has so far been extremely inefficient in dealing with waste. In short, humans are good miners but bad recyclers. Fortunately, there is plenty of room for improvement!

We could sidestep the recycling problem by moving to a more effective strategy: reusing. It means manufacturing objects that are intended to be used more than once. Think of the common disposable plastic bottle. If the bottles are separated in the waste collection stream, they can be melted and the plastic can be used to manufacture new bottles or other objects. This strategy is much

Electronic Waste and Rare Earths: Recycling the Needle in a Haystack

Rolf Jakobi

Practically all processes in industry and private consumption today are more or less dependent on electronic devices. Production, transportation, communication, energy, and food distribution would be unthinkable without electronic assistance. So the survival of our modern civilization is based on billions of electronic modules, chips, semiconductors, and lasers and on the uninterrupted and guaranteed supply of electricity. And practically all electronic devices contain very small portions of so-called rare earths, a classification comprising 17 metals, namely the 15 elements of the lanthanide group plus scandium and yttrium. For most applications in electronics, different materials cannot be substituted with the same efficiency, and that is why rare earths are said to be strategically important.

Rare earths are mined in only a few places in the world, where they can be found in high concentrations. According to the US Geological Survey, more than 97 percent of the active mines and 48 percent of the known reserves are located in China.[41] The minerals are often found in combination with other metals, and the processes of extraction and separation are complex and expensive. In the Western hemisphere, the Mountain Pass Rare Earths Mine in California was closed for economic and environmental reasons, though efforts have been made to reactivate its production. In fact, in order to reduce the Western dependence on supply from China, hasty attempts are being made worldwide to find new sources of minerals and reopen old mines.

While public and policy discussions about oil and gas depletion intensify, a much more imminent shortage, so far unnoticed but no less dangerous, is becoming visible—a shortage of not just rare earths but also other elements important in electronics, whose names probably are known only to a few experts. For example, a joint research project found that the reserves-to-production ratio for gallium, germanium, and indium is estimated to be less than two decades of supply.[42]

Officials in politics and industry have been slow to react, and the public has yet to seize the issue and form public opinion. Consequently, millions of tons of electronic equipment is not recycled in the appropriate way, and even worse, thousands of tons of important metals and other elements end up in uncontrolled waste deposits.

The critical nature of the rare earth situation became obvious when the Japanese coast guard arrested the captain of a Chinese fishing boat in 2010.[43] The Chinese government in turn stopped its supply of rare earths to Japan.[44] This event revealed two things: the Chinese know exactly where to hit foreign industries, and a country that owns strategic resources doesn't need strategic weapons anymore. Only a few weeks with no supply of rare earths would bring Western production to a standstill. In other words, in a trade war Western industries can be destroyed quickly and without a single shot. Many nations lack even a modest emergency stock of strategic materials such as rare earths, though the risky dependence on a single supplier has spurred various governments and industries to research solutions to the problem.

The rare earth issue has other thorny issues, too. Western authorities have long known that rare earths are mined in China under disastrous conditions for both miners and the environment. But, having decided to shut their own mines to avoid expensive investments in environmental protection, they seek the material from China at a discount price.

A considerable relaxation in the supply situation could be achieved if there were a functioning recycling process for electronic waste (e-waste). These strategic metals often exist in significantly higher concentrations in various products in circulation than they do in mines. A 2008 Eurostat survey of 28 European nations estimated e-waste at 1.8 million metric tons per year.[45] A US Environmental Protection Agency report estimated that 438 million new consumer electronic products were sold in the United States in 2009, while 181 million electronic products were discarded, 25 percent of which were collected for recycling.[46] However, recycling processes today are most often restricted to the recovery of iron, copper, aluminum, and glass. Some nations that boast exemplary recycling rates overall hide the fact that recycling rates of rare strategic elements are sometimes less than 1 percent.

Those who argue against rare earth recovery claim that recycling is too expensive. Indeed, only a few industrial plants in the world are technically capable of recycling rare earths. So it seems that practically all these elements meet their end in cement furnaces when the plastic boards containing them are incinerated. From there, they may end up in the slag, or bottom ash, that is combined with asphalt to cover streets. In other words, they are lost forever. Other, largely illegal ways of getting rid of e-waste are exports of so-called secondhand electronics to Asian or African countries. The Basel Convention (yet to be ratified by the United States) prohibits wealthy nations from exporting toxic waste to developing nations.[47] However, due to gaps in this legislation, declaring waste to be secondhand equipment makes a legal outlet possible, and authorities in overseas harbors do not have sufficient capacities to check the millions of containers and decide what is scrap and what is really still usable equipment. Facing this situation, many nations have established stricter e-waste recycling goals. However, an e-waste collection program will not necessarily recover all the strategic elements in electronics. Generally speaking, the current technology of recycling e-waste is far from state-of-the-art, having been borrowed from car-shredding technology, which is not well suited to this kind of waste. In addition, the recycling industry for precious and rare metals is often reluctant to share information, and this is an obstacle to the development of effective techniques and procedures.

Reliable supply and demand figures for rare earths and other strategic minerals are also hard to come by and, if available at all, show high variances. Information about proven or assumed resources is rare, and especially statistics from China and Russia are incomplete, inconsistent, and often kept secret.[48] Information about the amount in circulation in electronic products is also difficult to estimate. There are just a few milligrams within each device, but the products are distributed all over the world—in, for instance, about 340 million computers, 240 million TVs, and more than a billion cell phones in 2012.[49] To add to the confusion, different products have different life cycles, making it difficult to accurately project the material in circulation.

These uncertainties make it all the more important to recycle as much material as possible at the production end. Indium, for instance, is at present indispensable for touch screens and other displays, but it is likely that supplies will run out in the near future. We know little about the content of indium in different devices; estimates range from 40 to 260 milligrams per device.[50] But an indium recycling loop exists during production; only about 30 percent of the material that is processed ends up deposited on the target surface, and the rest is recovered.[51] Nevertheless, there is a yearly estimated drain of about 390 tons of this material, from the product side, and nobody knows where it goes. We have to ask ourselves how long we can afford such behavior.

There are several ways to improve the situation. The easiest one is at the technical level. All products, not only electronic equipment, have to be constructed so that they can be disassembled into separate modules, each still usable as a single entity. That means the products must be designed differently and robotic systems for rebuilding devices must be installed. It also means that prices for electronic equipment will rise sharply, and consumers should therefore make wiser choices about what they really need. We also need a sophisticated logistical system of collection as well as accurate indications about the chemical composition of various products, a task that has been the object of a European regulation.[52] Entropy inevitably increases along the way from the raw materials to the end products, so it must at least be possible to track the components along the value chain.

The recycling process itself also needs improvement. Chemists have long known that rare earths and noble metals can be separated. Therefore it is astonishing that so little effort has been undertaken to recycle more of the critical metals. After the commodities have been separated the residual proportion of e-waste can be dissolved in aqua regia (an acid) and then processed according to the prescriptions in the traditional chemical handbooks via ion exchange chromatography. Certainly, additional research and some upgrades will be needed to transfer the laboratory processes to an industrial scale. Creative solutions are already in sight. Japanese researchers have reported promising results in using microorganisms to separate rare earths through "bio-leaching."[53]

Product quality needs also to be improved so that products have a longer life cycle. Very often, only small modifications of essentially the same products motivate inflationary consumption. However, this step will be extremely difficult to implement, because it would reduce turnover and economic growth, shaking society's predominant economic beliefs to the core.

Additional systems for collecting e-waste can easily be established. Millions of small devices like mobile phones still end up in trash bins. Like soda bottle deposit programs, deposit programs for small electronic products can motivate consumers to return those products to retailers or manufacturers for recycling, but the sum paid at the time of purchase must be high enough that consumers remember it when they want to get rid of their old devices. In the long term, it might also be possible to switch to a leasing system that would require users to turn in their old device before getting a new one. Shifting away from the old way of thinking will require consumers to change their mind-set from that of an owner to that of a user.

A radical change is also necessary in our economic thinking. First, we need to stop the continuous chatter about growth. There is really nothing around us that continuously grows—aside from deadly cancer. That is a fact that everybody, without much effort, can observe. However, the apologists of growth in politics and economics adore perpetual growth like a holy cow.

Second, since all our resources are rare at some level, we need a circular-flow economy to preserve them for future use. That will demand a completely new economic approach—one that focuses on regeneration versus growth, availability versus profitability, value versus price, and energy versus money. After all, it is energy that we need to produce products, energy that we need to use them, and energy that we need to ultimately recycle them. So in reality it is not profitability that decides for or against recycling, it is energy and availability—including the energy of goodwill.

better than burning the bottles in incinerators or burying them in landfills, but it still takes energy to melt the bottles and make new ones. The best strategy of all is to reuse the bottles; it is a method that requires almost no energy, and it does not pose the downcycling problem.[54]

Reuse could be applied to a variety of industrial products, but it also requires a change in the way products are designed. For instance, in countries where reusing plastic bottles is not encouraged, they are often thin, to save materials, and that makes them easy to deform and nearly impossible to reuse. Instead, in countries like Germany where reuse is mandatory, plastic bottles are thicker and sturdier, to make them strong enough for the task. Aluminum beverage cans, on the other hand, are impossible to reuse the way they are designed today. Yet nothing would prevent someone from designing aluminum containers that could be reused several times, using for instance screw caps for sealing.

Hurdles for recycling come also from the "planned obsolescence" habit. Car bodies, for instance, are usually made in ordinary (that is, non-stainless) steel and are notorious for rusting easily. Despite modern surface treatments, rust remains a major factor in forcing owners to scrap an otherwise perfectly good car. Could we make car bodies using more durable materials such as stainless steel or titanium? Of course we could, but with the usual caveat: substituting takes energy. The energy required for making stainless steel is about twice that needed for ordinary steel, while titanium would require about 10 times as much. However, a car made in stainless steel or titanium would never rust and would last practically forever. That would lengthen the lifetime of the car and considerably reduce the amount of materials and energy needed. Unfortunately, however, this kind of strategy goes against the grain of every automotive industry norm. The market alone won't drive these industrial design changes without some legislative intervention. Designing products to be reusable and long lasting has never been popular. Although perceptions are slowly changing, for the vast majority reuse smacks of poverty. However, if an energy crisis hits us, we'll be forced to use the products we have for longer times, with all the limits and problems involved.

The problem is not theoretical but practical: it involves redesigning the world's industrial system, and that would entail enormous costs and long time frames. In practice, we cannot reasonably hope to match the expected speed of depletion in this way, but these measures could surely ease the problem, shifting it considerably toward a remote future.

The Efficiency Option

Most experts agree that adapting to resource scarcity will take a combination of better efficiency, better technologies, and better management. Especially where energy is concerned, efficiency is often considered the most crucial first response, leading to such measures as home insulation, compact or hybrid cars

FIGURE 7.3. A small, all-electric car. Electric cars are not just a substitute for the present, wasteful ways of transportation based on oversized and overweight cars. They are a completely new generation of small and efficient vehicles that may revolutionize the way we approach transportation.

for transportation, and high-efficiency lighting, among others. All these measures would allow us to live just as before, compensating for the higher costs of mineral commodities with a better efficiency in using them. It is, again, an attempt to win the Red Queen's race.

There is no doubt that energy efficiency is a good thing, at least in the short run. The problem is how well the concept can work in the long run to solve the depletion problem. One problem is that you need to invest money and resources in energy efficiency, which often leads to difficult choices. For instance, suppose you want to improve the efficiency of energy production by getting rid of an old and inefficient coal-fired plant. You could decide to build a more efficient gas-fired turbine plant. Alternatively, you could decide to invest in renewable energy. The renewable option would provide a clean break from harmful emissions, but it is also, at present, the most expensive. If you choose the gas-fired plant, you save the most money in the short term and obtain better efficiency and fewer emissions, but you also make an investment that locks the energy production system to fossil fuels for decades. A similar problem appears when you decide to get rid of your old gas-guzzling car. You could buy a compact car, a hybrid, or a full electric car. The least expensive choice is the compact, but it will lock you in to the use of gasoline for at least 10 years (the car's projected life expectancy). A fully electric car, on the contrary, will provide a clean break from the need of gasoline, but, right now, it will be more expensive.

To this problem we can add another: it turns out that the best energy efficiency is not necessarily obtained with the least use of rare and depletable materials. Take a look at what happened when we replaced the conventional light bulb, using tungsten filaments, with a new generation of light bulbs based on fluorescent light or on light-emitting diodes (LEDs). The new lights are much more efficient, but they use rare minerals that may soon be in short supply. Fluorescent lights need mercury in the mix of gases they contain, whereas LEDs use rare minerals such as gallium and indium. Is the replacement worthwhile, considering the challenges in recycling the rare materials contained in the new generation of lamps? And how about the harmful effects of mercury? In wealthy countries it is usually possible to have efficient systems for recycling old mercury-containing lamps, but that is not the case for poor countries. In practice, the mercury used for these lamps is shipped from rich countries to poor ones, where it is dumped somewhere, free to pollute the ecosystem and damage the health of local communities. Opinions vary on this point, but one thing is certain: energy can be produced in a sustainable way using renewables, but no technology based on rare materials such as mercury and gallium can be considered sustainable in the long run.

An even more general wrinkle in the efficiency approach appears in the form of the "Jevons paradox," what modern economists often describe as the "rebound effect." The concept is rather simple: if you are paying money for energy and you save some of it by being more efficient, then you have extra money to spend or invest. Most likely you'll invest this money in tasks that involve energy consumption. So in the end there will be no net decrease in the amount of energy you use. In other words, if you retrofit your oil-heated home to consume less fuel, you may decide to spend the money that you save on home heating on a vacation to Hawai'i, and the energy consumed by the plane will erase the savings.

A different approach to the same issues lies in consciously simplifying lifestyles. The varied movements forging this transformation go by many different names. But whether called "degrowth," transition, or just simple living, the goal at the nexus is to reduce resource consumption, individually and as a society. That may mean eating less energy-intensive foods, using public transportation instead of cars, avoiding plane travel and long-distance vacations, sharing big equipment (like cars) instead of owning it alone, and in general focusing social and business interests in a relatively small local area. While this may sound like downscaling, those who pursue it—whether in official Transition Towns or more casually—do so not just to curb consumption but also to improve their quality of life. Degrowthers, apparently, are not trying to win the Red Queen's race.

Their approach may solve at its root the Jevons paradox that plagues the concept of energy efficiency. Someone who is simply efficiency oriented may decide to save money by buying more efficient lightbulbs and then, because the cost of keeping them on is less, keep them on for longer times. On the contrary, a degrowther may turn the light off regardless of the energy price, using it only when truly needed and maybe using near-darkness to enjoy a romantic candlelight dinner.

Degrowth is not a complex strategy; it simply involves going with the flow. If the increased costs of mineral commodities make cars very expensive, you may use a bicycle—much less expensive. And if you can't even afford a bicycle, why not move closer to where you need to go and walk there? What we call "commuting" is the result of a historical phase when it was possible to allocate an enormous amount of resources to an inordinately expensive transportation system. That has led to a situation where people think it is normal to live at tens, or even hundreds, of miles from their workplace. But that has been normal only for the past century or so. If we can't afford to commute anymore, eventually we'll revert to our old ways, living close to our workplace and walking there—with little need for cars, trains, or other energy- and resource-expensive transportation systems.

Society can shed a lot of fat and still continue to function in a manner not unlike the present. The degrowth movement theorizes that living in a simpler society means being happier and living less stressful and more satisfactory lives. It could be, but we must also remember that losing weight too fast may be dangerous. In a spiral of reductions in the use of resources and energy, society could slowly (and perhaps not so slowly) lose its industrial base and revert to a purely agricultural society, as it was a few centuries ago. This would be the ultimate adaptation to mineral scarcity, but not exactly what most people would consider a positive result.

Even without going to such an extreme, though, the degrowth movement suffers somewhat of a public relations problem. It seems to have been accepted by only a tiny minority of the population of most countries. The mainstream media and general debates seem to ignore the concept completely. And in some ways degrowthers have become victims of the Jevons paradox, too: the resources they save are being used elsewhere, as there doesn't appear to exist, at present, a reduction in the use of natural resources in the world.

But it is also true that degrowth is rapidly ceasing to be a choice. Given the rapidly degrading economic conditions in many countries of the world, degrowth is becoming a forced condition. The long-term decline of the middle class in the United States has left a widening wage gap between employees and CEOs. The economic crisis that started with the financial collapse of 2008 left massive unemployment and underemployment in its wake in many countries of the world. Some places, such as China, still show robust economic growth (as of 2013), but several European countries seem to have taken a path of recession and decline, in particular the Mediterranean countries, such as Greece. It is too early to understand whether this path is a minor oscillation or truly irreversible—and whether it is going to spread to the whole world.

Forced degrowth is altogether different from intentional degrowth, and to get an idea of how it might play out we can look at the collapse of the Soviet Union in 1991, which was likely strongly affected by the increasing costs of exploiting mineral resources. As Dmitry Orlov describes in his book *Reinventing Collapse*, the life of Soviet citizens during and after the collapse involved a series of changes: reduction of life expectancy, increased rates of drug abuse and depression, increased incidence of illness, collapse of security due to rampant crime, the widening social gap between the rich and the poor, the decline and collapse of social services, and other factors that made it surely very unpleasant for those who had to live through it. Orlov maintains that the aftermath of economic collapse in Western countries might look very similar, and some hints that we are heading in that direction can already be seen nowadays.[55]

In the end, the question of adaptation is not just a technological one. It is perhaps much more a psychological one. The five stages of grief described by Elisabeth Kübler-Ross for people facing hard times or personal loss are well known: first there is denial, then anger, then bargaining, then depression, and finally acceptance.[56] This series may be too schematic to describe the nuances of people's real-world behavior, but it does have a logic and a certain degree of applicability. If we apply it to the behavior of society facing the problems generated by depletion, it is clear that, today, public opinion is mostly deeply grounded in the first stage: denial.

Many people have heard of peak oil or other bits and pieces of the depletion problem. The general reaction is to consider these concepts as extreme views held by Cassandras and catastrophists. Politicians everywhere seem to remain locked to the concept that growth is the only way to solve all problems. It is no coincidence that until just a few years ago several Western governments had implemented schemes called "cash for clunkers" in which car owners were rewarded with taxpayer money for junking perfectly good cars and buying new ones. In light of the increasing depletion problem of mineral resources, such schemes are truly madness, but they are the result of a view of the economic process that still neglects the fact that the economy strongly depends on the availability of low-cost mineral resources.

Not rarely, the reaction to depletion and the associated problems of pollution moves on to the second stage described by Kübler-Ross: anger. This reaction is most visible around the issue of climate change and most commonly takes the form of virulent, visceral attacks against the messenger. These attacks often involve not just individuals but entire sectors of society. Recently such attacks have become a widespread and accepted political stance of the American Republican right. Some business and political interests have tried to discredit individual scientists and the whole concept of scientific research and to launch anti-science propaganda campaigns.[57] Big money from big business plays a role in these efforts. However, the intensity of feelings seems to go beyond motives attributable to simple profit. And, of course, anger occurs on the other side as well, with fierce emotions stirred in those who understand climate change is happening but preventable.

The third stage of grief, bargaining, in essence, forms the basis of the "energy efficiency" concept—the idea that we can keep everything as it is if we are simply better at using what we have. This attitude is often accompanied by a remarkable faith in the power of science and technology. As for depression, the fourth stage, it is likely much more widespread than what can be seen through traditional or social media.

Only if we arrive at the fifth stage, acceptance, can we progress to adapting to the changes ahead. We need to accept that we cannot keep everything as it used to be in a world that is changing so deeply. In the end, we can't win the Red Queen's race.

The Shape of Things to Come

Debates about how to respond to mineral depletion tend to focus on the short term. But let's try to glimpse a little bit further into the future. We see that we are squeezed between the two complementary problems of resource depletion and ecosystem disruption. Together, these problems are making us become inhabitants of a new planet—one that will have a different climate and will be much poorer in terms of available resources. We also see that there is no obvious way of "solving" these problems as long as we intend to maintain the lifestyles that we have had until now.

Big changes are ahead, but exactly what kind?

Predictions about the future are always difficult to make. However, we can engage in a game of scenario building and try to see at least what kind of alternatives we might be facing. Jorgen Randers described a detailed four-decade scenario in his book *2052: A Global Forecast for the Next Forty Years*, in which he describes the possible evolution of a world squeezed between climate change and mineral depletion. Randers's conclusions are that the change may be gradual and involve some degree of adaptation but that, overall, we'll fail to react decisively and effectively against the tremendous difficulties we'll face in terms of climate change, overpopulation, and resource depletion. Randers's work is just an illustration of the many ways of seeing the future. Such attempts at predicting the future are fraught with difficulty, but they can be worthwhile if we don't attempt to make detailed predictions. Long-term trends, on the other hand, can be identified and analyzed.

Here, without establishing a specific time frame, we can look to a future in which the present problems will have played out their effects and the planet will have reached some kind of equilibrium after the great storm created by the industrial age. It could be a few centuries from now, or even much earlier than that.

First, it may be that one of the two problems (climate change and the associated ecosystem destruction) becomes so big and so intractable that it dominates the future. That might happen if the various enhancing feedbacks that govern the Earth's climate go out of control and shove the planet into a vastly different climatic state. One extreme is the so-called Venus scenario, as described, for instance, in James Hansen's book *Storms of My Grandchildren*.[58] In this case a runaway greenhouse effect would sterilize Earth, with temperatures at several hundred degrees Celsius and an atmosphere composed mainly of CO_2, like the planet Venus. Obviously humans couldn't survive that.

Even if the reality were less extreme, we could see a "post-Permian" climate situation, with tropical temperatures of about 50 to 60°C.[59] In such a case humans could survive only in the extreme northern and southern continental regions, in conditions completely different from the present ones. Humans might even find themselves inhabiting places such as a Greenland free of ice or even Antarctica.[60] We might find a way in such a future to adapt to these conditions. But in this scenario, the decline of humans would be so dramatic that our lifestyle may become similar to that of our ancestors of hundreds of thousands of years ago. Richard Duncan called this possibility the Olduvai scenario,[61] from the name of the African valley where fossils of our remote ancestors have been found.

There are elements that may make the Venus scenario physically impossible or at least extremely unlikely. The same cannot be said for the post-Permian scenario, since we know that such conditions did occur on the Earth in the past. However, although extreme warming scenarios cannot be ruled out, they are not necessarily our future. Ecosystem collapse is not unavoidable; it is a consequence of human actions. Even though we aren't currently acting to avoid it, we could choose to do something to mitigate the problem, or we could be forced to do so, although unwilling, by the depletion of all mineral resources and in particular of fossil fuels. So the consequences of climate change might not be so terrible as some extreme scenarios describe them.

Though we might be able to avoid the worst in terms of climate disruption, something that is truly unavoidable in our future is the disappearance of high-grade ores and the dispersal of the elements they contained all over the planet in forms that cannot be recovered—at least not without enormous energy costs. So what kind of future can we expect as the result of ore depletion?

FIGURE 7.4. The image of the Anthropocene: city lights, the result of energy and materials produced by mining. How long will they stay on?

One possibility is that we simply return to a purely agrarian society as the result of the disappearance of fossil fuels and the consequent disappearance of the energy needed to run an industrial society. After all, the world's economy was purely agrarian just a few centuries ago, and the big flaring up of fossil fuels could turn out to be just a short-lived episode—a peculiar moment of energy availability that generated a lot of commotion and movement but abated rapidly, returning humans to the condition that had been normal in the past ten thousand years or so. In 1976 Marion King Hubbert had already shown the world's fossil fuel consumption as a short-lived spike in a paper titled "Exponential Growth as a Transient Phenomenon in Human History."[62]

Again, this is a scenario that cannot be ruled out. A future agricultural civilization would have to cope with badly depleted soil resources left by the ruthless exploitation of a few centuries of the industrial age. But soil can reform, although it takes centuries, and such a civilization would eventually find a form of equilibrium, probably with a population much smaller than the present one. If it is any consolation, our descendants would not need the large amounts of resources that our society needs now. Just as people in the Middle Ages mined the remnants of Roman buildings to get iron and stone, our agrarian descendants would have plenty of metals from what we left: aluminum from our beverage cans, gold from our jewelry, copper from our pipes. They would also have plenty of iron and steel from our buildings and all other manner of stuff that we leave behind. Today we produce more than a billion tons of steel per year, but in Napoleon's time that figure was less than a million tons per year. Just using the iron we have produced and dispersed over the planet, our descendants could happily forge swords and plows (and perhaps also muskets and cannons) for tens of thousands of years.

Such a society would be poor compared to our standards. The surplus of energy produced by agriculture is small in comparison to what the opulent industrial society is accustomed to. In a purely agrarian society, the availability of charcoal, from scarce wood resources, would limit the ability to smelt metals, to build machinery, and to create all the structures that have made possible today's complex society. With these limits, this future society would be a low-technology system based mainly on human and animal labor. That wouldn't necessarily be so bad: after all, when Leonardo painted the *Mona Lisa* and Dante wrote *The Divine Comedy*, each was living in a purely agrarian society. But would such a society ever be able to restart an industrial revolution? Possibly not, since it would not have the same low-cost coal that started the industrial revolution a few centuries ago. But who can say? Maybe there are other ways to create a complex society.

It is also perfectly possible that, once we are reverted solely to agriculture, it will be forever. That raises the question of whether we will lose all the technological capabilities we have today. And since most of our high technology is based on electricity, that question of whether we'll be able to maintain a complex society like the present one boils down to a simple issue: whether or not we will be able to maintain the capacity to produce electric power. That doesn't mean that we need to maintain the same production capacity that we have now, just that we need to sustain electricity production for a long time, with power plants maintaining and renewing their power with power they themselves produce. In other words, the system must have a reasonably good energy efficiency and must not use rare and nonreplaceable minerals.

It is not impossible to attain these conditions. The production of electric power doesn't really need highly sophisticated equipment. After all, up to not long ago many countries produced most of their electricity by means of hydroelectric plants. This is not a complex technology, and it was perfectly possible to build these plants more than a century ago. We have also seen that technologies such as modern wind power, solar concentration, and photovoltaics have reasonably high EROEIs and can be manufactured without a critical need of rare materials. It may also be possible to return to presently abandoned nuclear technologies, such as the idea of breeding plutonium fuel from uranium or perhaps uranium from thorium. Breeder reactors, if they can be made to work with good efficiency, could lead to the availability of fuel for nuclear fission plants for at least a few hundred years. So there are ample possibilities for generating electric power without the need for fossil fuels. A future society could have electric power in relative abundance, even though not necessarily a steady flow of it if it were to be based on intermittent sources such as wind or sun. With this energy, it would be possible to maintain a certain flow of rare metals into the economy through careful management of the remaining mineral resources and by using a combination of recycling waste, reusing products, and substituting rare minerals with abundant ones. Closing the cycle of mineral resources is not impossible, though the abundance and low cost of old times will never come back.[63]

It is not easy to imagine the details of the society that will emerge on an Earth stripped of its mineral ores but still maintaining a high technological level. We can say, however, that most of the crucial technologies for our society can function without rare minerals or with very small amounts of them, although with modifications and at lower efficiency. If we can maintain a basic energy infrastructure for electric power, then, after the disappearance of fossil fuels, we can gradually rebuild the industrial system around materials abundant in the Earth's crust. The system would have to be less wasteful with resources and may be both slower and leaner, meaning that we won't be able to live at the mad speed of today, with the same rate of destruction of resources. We won't be able to maintain expensive and wasteful structures like highways and plane travel, but we might still keep the Internet, computers, robotics, long-range communications, public transportation, comfortable homes, food security, and more. We can hope the future won't lead us back to the times when peasants were condemned to a life of misery and physical exhaustion in the fields. It should be possible to use electricity, powered by renewables, to carry out many agricultural tasks that are today based on fossil fuels.[64]

The social structure of such a world is impossible to determine at present. We can only say that it will have to be much different from the present one. Surely running the Red Queen's race will take us somewhere; we just don't know where. We'll discover that as we move along the way.

CONCLUSION

A Mineral Eschatology

Eschatos is an ancient Greek term meaning "last," and the term "eschatology" has been used to refer to the ultimate end of the world and of humankind, a notion typically reserved for religious or philosophical studies. In recent times a new form of eschatology has appeared—physical eschatology.[1] Studies in this field investigate how the Earth will end as the sun evolves, or how the universe will end as it expands. Typically, the time scales of these natural processes can be projected in billions of years. However, if we define "eschatological" as meaning "occurring on a grand scale" and, at the same time, "irreversible," we see that we are facing a true mineral eschatology that may occur within the lifetime of most of the people living today.

In the past few centuries, the Earth saw a gigantic chemical reaction launched by the burning of carbon that had been buried in the crust for hundreds of millions of years. The reaction picked up speed and burned in a more and more intense fire. We are, perhaps, at the peak of this immense fire, and perhaps we are starting to see it showing signs of decline. Like all fires, this immense chemical reaction devours its fuel, and in the end it will flicker out.

As the great fossil fire fades, all the other mineral resources that the planet had accumulated over time are also disappearing in the form of the highly concentrated ores that have been used to build our society. One day in the future, without veins, without wells, without ores, we'll see the disappearance of the mining machines, of the drilling rigs, of the offshore platforms. We'll see the disappearance of the very concept of mines, holes dug deep underground to recover the precious minerals that the planet accumulated for us long ago. Miners will disappear, too, with their picks, their helmets, their lights, and their dirty faces.

It is the end of a cycle that, in geological terms, was extremely short but that for us seemed to be the way things were to be forever and ever. It wasn't so; it was only the brief cycle of the period we call Anthropocene, where humans thought themselves masters of a whole planet. But the planet was plundered to the utmost limit, and what we will be left with are only the ashes of a gigantic fire. We are leaving to our descendants a heavy legacy in terms of radioactive waste, heavy metals dispersed all over the planet, and greenhouse gases—mainly CO_2—accumulated in the atmosphere and absorbed in the oceans. The Earth will never be the same; it is being transformed into a new and different planet.[2]

It appears that we found a way to travel to another planet without the need for building spaceships. It is not obvious that we'll like the place, but there is no way back; we'll have to adapt to the new conditions. It will not be easy, and we can speculate that it will lead to the collapse of the structure we call civilization, or even the extinction of the human species. But neither is unavoidable.

By using solar energy and technologies that don't require rare and exhaustible elements, we are perfectly able, in principle, to build a society that manages flows of energy comparable to those we manage today. Closing the cycle of rare minerals is not impossible if we learn to use much less. We can create a society able to use this energy to keep a reduced supply of minerals sufficient to maintain an industrial infrastructure. That society would have to be extremely careful to avoid wasting its precious resources, and it would see some of our habits—air travel by jet planes, for instance—as dangerous extravagances. But such a society could maintain our technological level and improve it. It could engage in the exploration of space, in fundamental research, in the development of artificial intelligence, in all forms of art, and in other human pursuits that can't be conceived without the prosperity that comes from a significant supply of energy and materials.

Using this supply we can manage to maintain—and keep increasing—the knowledge that we have accumulated in the past millennia. We can use this knowledge to remedy the damage that we caused to the planet's ecosystem and return it to the condition it was in when we inherited it—a planet rich in life and diversity, a condition that we can maintain for millennia, or longer.

AFTERWORD

We Can Stop Plundering the Planet:

An Earth Citizen's View of Modern Mining

Karl Wagner

There have always been huge mineral exploitation sites visible to the naked eye: coal mines gouging the earth's surface; gigantic trucks circling toward the bottom of gold, copper, and other mines on what seem to be many-mile-long spiral staircases; colossal mining machines upturning entire coastlines in the search for diamonds. However, apart from these extremes, mining has been a rather spatially concentrated effort with much of the activity underground, and therefore out of view.

That is about to change. We are in the middle of a new phase of large-scale mineral exploitation, which not only destroys large landscapes but also poses a larger than ever danger for massive and lasting pollution affecting nature, wildlife, and humans. And some of the most extreme examples can be found in the fossil-fuel arena.

It seems that humanity is determined to leave as few mineral resources as possible for future generations. A treasure, generated in millions—even billions—of years, is being blasted or drilled away in about 200 years in a take-no-prisoners approach.

The defining factor in the extraction extravaganza unfolding before us is the move by oil and gas companies to exploit unconventional fossil fuels such as shale gas, heavy oil, deep-sea oil, or bitumen. That move has represented a big, unexpected shift in thinking about how long fossil fuels might prevail in our lives. In 2001, at the beginning of what was to be a historical ramping up of petroleum prices, BP adopted its "Beyond Petroleum" slogan. It was seen as a step toward an economy and a society based on solar energy, ultimately ending the era of fossil fuels and their problems of pollution and depletion. But that was not to be, and instead, the opposite happened: as the price of oil rose, companies found that they could reap enormous, short-term profits by exploiting resources that, before, had been too expensive to develop.

So "Beyond Petroleum" has turned into something like "Blast the Planet," and the rush to exploit unconventional oil and gas resources is on, stirring great hopes for a new age of fossil fuels. In turn, this rush is providing the energy resources needed to maintain the production of most mineral commodities despite their gradual depletion. Humankind seems intent to party as long as possible; but nothing can last forever, and there is sure to be a mighty hangover, once the party is over.

Unfortunately, as the need to switch to clean and renewable energy sources and limit the exploitation of mineral resources becomes more and more evident, the general attitude becomes more and more locked into old ways of thinking—helped greatly by the corporate spin machine. The messages that machine is passing to the public and to policy makers range from "there is no peak oil" to "fossil fuels exist in abundance" to "we can continue with our lifestyle." What they leave out are two messages essential to human well-being on this planet: the more we extract, the faster we run out of resources; and if we keep burning fossil fuels, we risk crossing the threshold to nonlinear, out-of-control climate change. The strategy chosen by the oil industry is going to lock us into the fossil-fuel economy longer than necessary, wasting some extremely precious years—not to mention financial resources that are desperately needed for investments into a solar future, before climate change becomes impossible to deal with effectively.

There seems to be no mechanism within the current economic system that could lead oil companies—or any large company—to adequately value long-term benefits for everyone, including, ultimately, themselves. Big corporations exist to optimize their (short-term) profits; it is what they were created for. But the damage done by companies seeking profit needs to be controlled by governments that represent the people and their well-being. Unfortunately, it seems that, today, politics has lost its way and has become a pressure-group-servicing sector devoid of any vision for the common good or the planet. Politicians are trying to convey an image of themselves doing their utmost (or at least something) to save humanity from the worst impacts of climate change, but at the same time they are handing out permits to further increase the exploitation of ever more exotic and damaging mining operations.

History shows us that remarkable results can come from major international protocols that limit environmental damage. Consider the ban on above-ground nuclear detonations, thanks to a treaty signed in 1963. Or the1987 Montreal Protocol on Substances that Deplete the Ozone Layer, designed to phase out chlorofluorocarbons, or CFCs. Or, particularly pertinent to the fossil-fuel issue, the Kyoto Protocol, the climate-change treaty—adopted in 1997 and put into effect in 2005—that slowly but gradually influences climate-change policies in more and more countries around the globe.

Many citizens around the world are concerned by the developments in the mining sector. Countless local initiatives have already sprung up to fight fracking, mountaintop removal, Arctic drilling, and the exploitation of tar sands as concerns about landscape destruction, chemical pollution, and climate impact mount, and as the economic value of these operations is increasingly questioned. But what seems to be missing is a global response.

So far, no major international treaty has dealt with directly limiting the exploitation of mineral resources worldwide. However, in 2003 the Oil Depletion Protocol (also known as the Rimini Protocol) was proposed by oil expert Colin Campbell and would have placed a cap on the quantity of oil that could be produced.[1] It was an innovative idea that was, in a way, the "mirror" of the Kyoto agreement. While the Kyoto Protocol aimed to reduce emissions from burning fossil fuels in order to fight global warming, the oil protocol aimed directly at fighting the fossil-fuel depletion problem. The idea was to reduce extraction rates so that some resources would be left for future generations, especially for those who couldn't afford the expensive oil that remains after the cheap resources have been destroyed. Although Campbell's proposed protocol was aimed only at oil, it could have been the first step toward limiting the extraction of rare mineral resources with future generations in mind. Unfortunately, however, the oil protocol went against the grain of current economic thought, which assumes that growth and free markets can solve all problems. So, after some initial interest, the idea was abandoned.

That doesn't mean, however, that the public has lost interest in limiting mining through policy actions. Quite the opposite, it seems. For instance, the issue of drilling for oil in the Arctic National Wildlife Refuge has been an ongoing political controversy in the United States since 1977, and a European ban on mining in the Arctic was recently proposed.[2] Even though it was rejected after intensive lobbying by the oil industry, it is clear that a drive exists, especially within the European Union, to limit mining and its associated damage.[3] In general, there is widespread agreement that some regions of our planet—like national parks, wildlife refuges, and world heritage sites—are too important, environmentally or culturally, to be opened to mining. An entire continent, Antarctica, has been declared off limits to mining, according to the Antarctic Treaty System. And even in nonremote regions, politics can be sensitive to issues related to human well-being. For instance, the European Union has recently stepped back from a policy that would have significantly escalated the use of automotive biofuels, which in turn would have had a negative impact on food production and biodiversity.[4]

So special interest groups and their lobbyists, after all, are not all powerful. It is possible to enact legislative action to limit the damage done by mining and lengthen the duration of the remaining reserves. The battlefield, however, is political, and politics is becoming more and more a question of managing the media in what is called "consensus building." So far, the lobbies associated with the oil industry and the mining industry have been successful in thwarting most attempts to enact effective legislation. In the fight against global warming, they have run spin campaigns aimed at discrediting either the science behind the proposed measures or the individual scientists engaged in research in the relevant scientific fields.[5] These campaigns have been successful in that all attempts to agree on worldwide measures to fight global warming appear to have stalled, at present, just as the attempts to free humankind from the slavery of fossil fuels also seem to have stalled.

To succeed we need to understand the rules of the game. We need an overarching, unifying perspective, a global target, that can unite the aims of the many concerned citizens and local initiatives. We need a communications strategy that can effectively contrast the effect of spin campaigns on the public and on policy makers. We need to make the information produced by scientific research accessible—and understandable—to anyone, not just other experts in specific fields.

We also need to help and support grassroots movements that can influence policy makers. By and large, they are local or regional and have to fight battles over fracking, mountaintop removal, or other destructive extraction practices on their own. The wider public—outside of affected areas— has not yet fully grasped the magnitude of this new phase of mineral exploitation. The institutionalized conservation movement (especially outside the United States and Canada) hasn't, either. Still, the awareness that it is necessary to fight back the spin campaigns of the oil industry is growing and will hopefully lead to networking efforts, linking the many local groups with each other and with larger entities. Turning this growing awareness into real change on the ground will also require practical initiatives that are likely to gain traction in world politics and provide unifying targets.

It might be time, now, to revive and expand the idea behind the Rimini Protocol to place a legal cap on the exploitation of rare nonrenewable resources. Such a cap would not only provide a formidable tool to avoid the worst effects of climate change, it could also change our economic and industrial practices and lead us from a flow-through economy, where we see products more or less passing by on their way from the mine to the landfill, toward a circular economy, where materials and especially nonrenewable resources have cycles of usage. The focus would then shift from new production toward repair, maintenance, and resource efficiency, undoubtedly producing a great number of meaningful and local jobs along the way.

These efforts are our best hope for the future. If we fail, we will remain locked to fossil fuels until we can no longer extract them and climate destruction wreaks havoc on the Earth. If we succeed, we can fully embrace the energy source that is abundant, reliable, and always there—the sun.

Acknowledgments

This book has benefited from the support, encouragement, suggestions, corrections, and materials provided by a group of people deserving heartfelt thanks from the author: Colin Campbell, Stefano Caporali, Toufic El Asmar, Suren Erkman, Steven Featherstone, Jacques Grinewald, Ian Johnson, Rembrandt Koppelaar, Jean Laherrere, Alessandro Lavacchi, Dennis Meadows, Magne Myrtveit, Massimo Nicolazzi, Marco Pagani, Joni Praded, Jorgen Randers, Fabrizio Sibilla, Stuart Staniford, Karl Wagner, and Leigh Yaxley.

Notes

Preface

1. D. Meadows, D. Meadows, J. Randers, and W. W. Beherens III, *The Limits to Growth*, 2nd ed. (New York: Universe Books, 1974).
2. U. Bardi, *Revisiting the Limits to Growth* (New York: Springer, 2011); and G. M. Turner, "Updated Comparison of The Limits to Growth with Historical Data," *GAIA—Ecological Perspectives for Science and Society* 21, no. 2 (2012): 116–24.
3. G. M. Turner, "A Comparison of the Limits to Growth with Thirty Years of Reality," Socio-Economics and the Environment in Discussion (SEED) working paper 19 (Canberra, Australia: CSIRO Sustainable Ecosystems, June 2008).
4. A. Diederen, *Global Resource Depletion* (Amsterdam: Eburon, 2010); P. Bihouix and B. de Guillebon, *Quel futur pour les métaux?* (Paris: EDP Sciences, 2010); U. Bardi and M. Pagani, "Peak Minerals," guest post on the Oil Drum website, October 15, 2007, http://www.theoildrum.com/node/3086; U. Bardi, "The Universal Mining Machine," post on the Oil Drum website, January 24, 2008, http://www.theoildrum.com/node/3451; R. L. Moss, E. Tzimas, H. Kara, P. Willis, and J. Kooroshy, *Critical Metals in Strategic Energy Technologies*, JRC Scientific and Technical Reports (Luxembourg: European Commission Joint Research Centre Institute for Energy and Transport, 2011), accessed via the Scribd website, http://www.scribd.com/doc/71562416/111031-Rare-Earth-EUreport-Critical-Metals-in-Set; C. Clugston, *Scarcity: Humanity's Final Chapter?* (Booklocker.com, Inc., 2012); and A. M. Bradshaw and T. Hamacher, "Nonregenerative Natural Resources in a Sustainable System of Energy Supply," *ChemSusChem* 5, no. 3 (2012): 550–62.

Chapter One: Gaia's Gift

1. U. Bardi, *Il Libro della Chimera* (Rome: Editori Riuniti, 2001).
2. "The Descent of the Goddess Ishtar in the Underworld," from *The Civilization of Babylonia and Assyria*, by M. Jastrow (1915), Internet Sacred Text Archive, http://www.sacred-texts.com/ane/ishtar.htm.
3. A. Wegener, "Die Herausbildung der Grossformen der Erdrinde (Kontinente und Ozeane), auf geophysikalischer Grundlage," in *Petermanns Geographische Mitteilungen* 63 (1912): 185–95, 253–56, 305–9, presented at the annual meeting of the German Geological Society, Frankfurt, Germany, January 6, 1912.
4. P. D. Ward, *The Medea Hypothesis: Is Life on Earth Ultimately Self-Destructive?* (Princeton, N.J.: Princeton University Press, 2009).
5. T. Tyrrell, *On Gaia: A Critical Investigation of the Relationship between Life and Earth* (Princeton, N.J.: Princeton University Press, 2013).
6. R. Branch, "Goddess Worship," *Profile* (published by the Watchman Fellowship, Inc.), 1994, http://www.watchman.org/profiles/pdf/goddessprofile.pdf.

7. Burning Man, accessed November 20, 2012, http://www.burningman.com.
8. C. Frankel, *Volcanoes of the Solar System* (Cambridge: Cambridge University Press, 1996).
9. T. Kleine, "Geoscience: Earth's Patchy Late Veneer," *Nature* 477, no. 7363 (2011): 168–69.
10. R. Lindsey, "Ancient Crystals Suggest Earlier Ocean," NASA Earth Observatory (online), March 1, 2006, http://earthobservatory.nasa.gov/Features/Zircon/.
11. G. Wächtershäuser, "Before Enzymes and Templates: Theory of Surface Metabolism," *Microbiological Reviews* 52, no. 4 (1988): 452–84.
12. C. Sagan and G. Mullen, "Earth and Mars: Evolution of Atmospheres and Surface Temperatures," *Science* 177, no. 4043 (1972): 52–56.
13. M. T. Rosing, D. K. Bird, N. H. Sleep, and C. J. Bjerrum, "No Climate Paradox under the Faint Early Sun," *Nature* 464, no. 7289 (2010): 744–47.
14. R. L. Armstrong, "Radiogenic Isotopes: The Case for Crustal Recycling on a Near-Steady-State No-Continental-Growth Earth," *Philosophical Transactions of the Royal Society A* 301, no. 1461 (1981): 443–72.
15. C. Bounama, S. Franck, and W. von Bloh, "The Fate of Earth's Ocean," *Hydrology and Earth System Sciences* 5, no. 4 (2001): 569–75.
16. J. F. Kasting and M. Tazewell Howard, "Atmospheric Composition and Climate on the Early Earth," *Philosophical Transactions of the Royal Society B* 361, no. 1474 (2006): 1733–42.
17. P. Cloud, "Paleoecological Significance of the Banded Iron-Formation," *Economic Geology* 68, 7 (1973): 1135.
18. G. Price, P. J. Valdes, and B. W. Sellwood, "A Comparison of GCM Simulated Cretaceous 'Greenhouse' and 'Icehouse' Climates: Implications for the Sedimentary Record," *Palaeogeography, Palaeoclimatology, Palaeoecology* 142, no. 3–4 (1998): 123–38.
19. S. Franck, C. Bounama, and W. von Bloh, "Causes and Timing of Future Biosphere Extinction," *Biogeosciences Discussions* 2 (2005): 1665–79.
20. The KamLAND Collaboration, "Partial Radiogenic Heat Model for Earth Revealed by Geoneutrino Measurements," *Nature Geoscience* 4 (2011): 647–51.
21. L. R. Kump, S. L. Brantley, and M. A. Arthur, "Chemical Weathering, Atmospheric CO_2, and Climate," *Annual Review of Earth and Planetary Sciences* 28 (2000): 611–67, DOI: 10.1146/annurev.earth.28.1.611.
22. D. Raup, and J. Sepkoski Jr., "Mass Extinctions in the Marine Fossil Record," *Science* 215, no. 4539 (1982): 1501–3; and J. Alroy, "Dynamics of Origination and Extinction in the Marine Fossil Record," *Proceedings of the National Academy of Sciences of the United States of America* 105, suppl. 1 (2008): 11536–42.
23. L. W. Alvarez, W. Alvarez, F. Asaro, and H. V. Michel, "Extraterrestrial Cause for the Cretaceous–Tertiary Extinction," *Science* 208, no. 4448 (1980): 1095–108.
24. G. Keller, "The Cretaceous–Tertiary Mass Extinction, Chicxulub Impact, and Deccan Volcanism," in *Earth and Life*, ed. J. A. Talent (Springer Science+Business Media B.V., 2011), 759–93, doi:10.1007/978-90-481-3428-1_25.

25. P. Cloud, "Paleoecological Significance of the Banded Iron-Formation," *Economic Geology* 68, no. 7 (1973): 1135.
26. J. M. Robinson, "Lignin, Land Plants, and Fungi: Biological Evolution Affecting Phanerozoic Oxygen Balance," *Geology* 18, no. 7 (1990): 607–10.
27. D. H. Meadows, D. L. Meadows, J. Randers, and W. W. Behrens III, *The Limits to Growth* (New York: Universe Books, 1972).
28. C. Campbell., *The Coming Oil Crisis* (Essex, UK: Multi-Science Publishing, 2004).
29. It is promising that a large number of books and academic papers are now being written on natural limits. Many are listed in two recent books by Colin J. Campbell: C. Campbell, *Campbell's Atlas of Oil and Gas Depletion*, 2nd ed. (New York: Springer, 2013); and C. Campbell, ed., *Peak Oil Personalities* (Skibbereen, Ireland: Inspire Books, 2011).
30. H. D. Klemme and G. F. Ulmishek, "Effective Petroleum Source Rocks of the World: Stratigraphic Distribution and Controlling Depositional Factors," Search and Discovery Article 30003 (1999), condensed and adapted from the article published in *AAPG Bulletin* 75 (1991): 1809–51, http://www.searchanddiscovery.com/documents/Animator/klemme2.htm.
31. J. Laherrere, "No Free Lunch, Part 1: A Critique of Thomas Gold's Claims for Abiotic Oil," FromTheWilderness.com, 2004, http://www.fromthewilderness.com/free/ww3/102104_no_free_pt1.shtml.
32. M. Höök, U. Bardi, L. Feng, and X. Pang, "Development of Oil Formation Theories and Their Importance for Peak Oil," *Marine and Petroleum Geology* 27, no. 9 (2010): 1995–2004.

Chapter Two: Plundering the Planet

1. M. Schirber, "The Chemistry of Life: The Human Body," Livescience (online), April 16, 2009, http://www.livescience.com/3505-chemistry-life-human-body.html.
2. C. B. Field, M. J. Behrenfeld, J. T. Randerson, and P. Falkowski, "Primary Production of the Biosphere: Integrating Terrestrial and Oceanic Components," *Science* 281, no. 5374 (1998): 237–40.
3. H. W. Kelley, *Keeping the Land Alive: Soil Erosion—Its Causes and Cures*, FAO Soil Bulletin 50 (Rome: FAO, 1983).
4. S. C. Hodges, "Basic Concepts," chapter 1 in *Soil Fertility Basics: NC Certified Crop Advisor Training* (North Carolina State University Soil Science Extension, n.d.).
5. F. H. Beinroth, H. Eswaran, P. F. Reich, and E. Van Den Berg, "Land Related Stresses in Agroecosystems," in *Stressed Ecosystems and Sustainable Agriculture*, eds. S. M. Virmani, J. C. Katyal, H. Eswaran, and I. P. Abrol (New Delhi: Oxford and IBH, 1994).
6. F. R. Troeh, J. A. Hobbs, and R. L. Donahue, *Soil and Water Conservation* (Upper Saddle River, N.J.: Prentice Hall, 1999).

7. Food and Agriculture Organization of the United Nations, *The State of the World's Land and Water Resources for Food and Agriculture: Managing Systems at Risk* (Rome: FAO, 2011).
8. O. Cerdan, G. Govers, Y. Le Bissonnais, K. Van Oost, J. Poesen, N. Saby, A. Gobin, A. Vacca, J. Quinton, K. Auerswald, A. Klik, F. J. P. M. Kwaad, D. Raclot, I. Ionita, J. Rejman, S. Rousseva, T. Muxart, M. J. Roxo, and T. Dostal, "Rates and Spatial Variations of Soil Erosion in Europe: A Study Based on Erosion Plot Data," *Geomorphology* 122, no.1–2 (2010): 167–77.
9. A. Freedman, "Causes of Midwest Drought: La Nina and Global Warming Thought to Contribute to Dry Weather," *Huffington Post*, July 21, 2012, http://www.huffingtonpost.com/2012/07/21/causes-of-midwest-drought-2012_n_1690717.html.
10. Earth Policy Institute, "Full Planet, Empty Plates: The New Geopolitics of Food Scarcity," press release promoting the book of the same name by Lester R. Brown, September 27, 2012, http://www.earth-policy.org/books/fpep/press_release_fp.
11. H. Eswaran, R. Lal, and P. F. Reich, "Land Degradation: An Overview," in *Responses to Land Degradation*, ed. E. M. Bridges, I. D. Hannam, L. R. Oldeman, F. W. T. Pening de Vries, S. J. Scherr, and S. Sompatpanit (New Delhi: Oxford Press, 2001), http://www.nrcs.usda.gov/wps/portal/nrcs/detail/soils/use/?cid=nrcs142p2_054028.
12. Food and Agriculture Organization of the United Nations, *Save and Grow: A Policymaker's Guide to the Sustainable Intensification of Smallholder Crop Production* (Rome: FAO, 2011).
13. U. Bardi, T. El Asmar, and A. Lavacchi, "Turning Electricity into Food: The Role of Renewable Energy in the Future of Agriculture," *Journal of Cleaner Production* 53, no. 15 (2013): 224–31.
14. "The Neolithic Flint Mines of Sussex: Britain's Earliest Monuments," Bournemouth University Archaeology Group, accessed August 2011, http://www.bournemouth.ac.uk/caah/landscapeandtownscapearchaeology/neolithic_flint_mines_of_sussex.html.
15. A. W. Cramb, "A Short History of Metals," Carnegie Mellon University Department of Materials Science & Engineering (online), http://neon.mems.cmu.edu/cramb/Processing/history.html, accessed November 7, 2011.
16. M. Cotterell, *The Terracotta Warriors: The Secret Codes of the Emperor's Army* (Rochester, Vt.: Bear and Company, 2004), 102.
17. S. P. M. Harrington, "Behind the Mask of Agamemnon," *Archaeology* 52, no. 4 (1999), http://www.archaeology.org/9907/etc/mask.html.
18. M. Stanczak, "A Brief History of Copper," CSA/Proquest, October 2005, http://www.csa.com/discoveryguides/copper/overview.php.
19. J. A. J. Gowlett, *Ascent to Civilization: The Archeaology of Early Humans*, 2nd ed. (New York: McGraw-Hill, 1992).

20. Waheenee, *Waheenee: An Indian Girl's Story Told by Herself to Gilbert L. Wilson* (1921; reprint, Lincoln: University of Nebraska Press, 1981), as cited in D. K. Jordan, "Ancient Metallurgy: A Beginner's Guide for College Students," December 2, 2007, http://weber.ucsd.edu/~dkjordan/arch/metallurgy.html.
21. J. Grout, "Lead Poisoning and Rome," Encyclopaedia Romana, accessed February 22, 2012, http://penelope.uchicago.edu/~grout/encyclopaedia_romana/wine/leadpoisoning.html.
22. J. Martín-Gil, F. J. Martín-Gil, G. Delibes-de-Castro, P. Zapatero-Magdaleno, and F. J. Sarabia-Herrero, "The First Known Use of Vermillion," *Experientia* 51, no. 8 (1995): 759–61.
23. C. E. Cramer, "What Caused The Iron Age?" paper for History 303 (university unnamed), December 10, 1995, http://www.claytoncramer.com/unpublished/Iron2.pdf.
24. S. Srinivasan and S. Ranganathan, "Wootz Steel: An Advanced Material of the Ancient World," Indian Institute of Science (Bangalore), http://materials.iisc.ernet.in/~wootz/heritage/WOOTZ.htm, accessed June 27, 2012.
25. R. Solnit, "Winged Mercury and the Golden Calf," *Orion*, September/October 2006, http://www.orionmagazine.org/index.php/articles/article/176/.
26. "Timeline of Chemical Elements Discoveries," *Wikipedia*, accessed July 24, 2012, http://www.en.wikipedia.org/wiki/Timeline_of_chemical_elements_discoveries.
27. W. S. Jevons, *The Coal Question: An Inquiry Concerning the Progress of the Nation, and the Probable Exhaustion of Our Coal Mines* (London: McMillan and Co., 1865).
28. A. Hochschild, *King Leopold's Ghost: A Story of Greed, Terror, and Heroism in Colonial Africa* (Boston: Mariner Books, 1998).
29. M. K. Hubbert, "Nuclear Energy and the Fossil Fuels," report presented before the Spring Meeting of the Southern District, American Petroleum Institute, Plaza Hotel, San Antonio, Texas, March 7–9, 1956.
30. P. Wack, "Scenarios: Uncharted Waters Ahead," *Harvard Business Review*, September 1985.
31. C. J. Campbell and J. H. Laherrere, "The End of Cheap Oil," *Scientific American*, March 1998.
32. U. Bardi, "No Peak Oil Yet? The Limits of the Hubbert Model," post on the Oil Drum website, January 17, 2011, http://www.theoildrum.com/node/7241.
33. S. F. Hayward, "The Gas Revolution," *The Weekly Standard* 16, no. 30 (2011), http://www.weeklystandard.com/articles/gas-revolution_557014.html.
34. "Nuclear Arms Race," *Wikipedia*, accessed August 2011, http://www.en.wikipedia.org/wiki/Nuclear_arms_race.
35. C. T. Montgomery and M. R. Smith, "Hydraulic Fracturing: History of an Enduring Technology," *Journal of Petroleum Technology*, December 2010, http://www.spe.org/jpt/print/archives/2010/12/10Hydraulic.pdf.

36. B. Faucon and K. Johnson, "US Redraws World Oil Map," *Wall Street Journal*, November 13, 2012, http://online.wsj.com/article/SB10001424127887324073504578115152144093088.html.
37. J. D. Hughes, "Drill, Baby, Drill: Can Unconventional Fuels Usher in a New Era of Energy Abundance?" Post Carbon Institute, February 2013, http://www.postcarbon.org/drill-baby-drill; and R. T. Pierrehumbert, "The Myth of 'Saudi America,'" *Slate*, February 6, 2013, http://www.slate.com/articles/health_and_science/science/2013/02/u_s_shale_oil_are_we_headed_to_a_new_era_of_oil_abundance.html; A. Berman, "The Big Deal about US Energy Self-Sufficiency,", post on the Oil Drum website, October 31, 2012, http://www.theoildrum.com/pdf/theoildrum_9584.pdf.
38. M. J. Economides and K. G. Nolte, *Reservoir Stimulation*, 3rd ed. (West Sussex, England: John Wiley & Sons, 2000).
39. Hughes, "Drill, Baby, Drill."
40. E. Cantarow, "Meet Anthony Ingraffea—From Industry Insider to Implacable Fracking Opponent," EcoWatch, January 2, 2013, http://ecowatch.com/2013/industry-insider-to-fracking-opponent/.
41. M. Fischetti, "Fracking Safety: Scientific Truths Are Emerging," *Scientific American*, April 20, 2012, http://www.scientificamerican.com/article.cfm?id=fracking-evolving-truth-natural-gas.
42. C. W. Abdalla and J. R. Drohan, "Water Withdrawals for Development of Marcellus Shale Gas in Pennsylvania: Introduction to Pennsylvania's Water Resources" (University Park: The Pennsylvania State University, 2008), http://pubs.cas.psu.edu/FreePubs/pdfs/ua460.pdf.
43. M. Meinshausen, N. Meinshausen, W. Hare, S. C. B. Raper, K. Frieler, R. Knutti, D. J. Frame, and M. R. Allen, "Greenhouse-Gas Emission Targets for Limiting Global Warming to 2°C," *Nature* 458 (2009): 1158–62, http://www.nature.com/nature/journal/v458/n7242/abs/nature08017.html, plus the author's adjustment for <350ppm CO_2 target.
44. F. Harvey, "Natural Gas Is No Climate Change 'Panacea,' Warns IEA," *The Guardian*, June 6, 2011, http://www.theguardian.com/environment/2011/jun/06/natural-gas-climate-change-no-panacea.
45. "Golden Rules for a Golden Age of Gas," *World Energy Outlook* series of the International Energy Agency, May 29, 2012, http://www.worldenergyoutlook.org/goldenrules/.
46. E. Mearns, "Unconventional Oil and Gas: A Game Changer?" PowerPoint presentation for the 10th Annual ASPO Conference, Vienna, May 30 through June 1, 2012, http://www.aspo2012.at/wp-content/uploads/2012/06/Mearns_aspo2012.pdf; and D. J. Murphy and C. A. S. Hall, "Energy Return on Investment, Peak Oil, and the End of Economic Growth," *Annals of the New York Academy of Sciences* 1219 (2011): 52–72, abstract, http://www.ncbi.nlm.nih.gov/pubmed/21332492.

47. A. D. Rossin, "US Policy on Spent Fuel Reprocessing: The Issues," PBS *Frontline*, n.d., accessed August 2011, http://www.pbs.org/wgbh/pages/frontline/shows/reaction/readings/rossin.html.
48. R. E: Langford, *Introduction to Weapons of Mass Destruction: Radiological, Chemical, and Biological* (Hoboken, N.J.: John Wiley & Sons, 2004).
49. "World Uranium Mining Production," World Nuclear Association, accessed March 31, 2012, http://www.world-nuclear.org/info/inf23.html.
50. Data about electric energy production are taken from *World Energy Outlook 2013* International Energy Agency, http://www.worldenergyoutlook.org/publications/weo-2013/ and "Monthly Electricity Statistics, Sep 2013," International Energy Agency, 2013, http://www.iea.org/stats/surveys/mes.pdf; data about the world nuclear reactors are available at the IAEA-PRIS Database on Nuclear Power Reactors at http://www.iaea.org/programmes/a2 (accessed December 23, 2013).
51. Uranium demand under the three WNA future scenarios can be found in "Uranium Markets," World Nuclear Association, updated June 2013, http://www.world-nuclear.org/info/inf22.html. The press release for the publication of the 2009 edition of the Red Book contains a warning statement about uranium shortages; see "Latest Data Shows Long-Term Security of Uranium Supply," Nuclear Energy Agency, July 20, 2010, http://www.nea.fr/press/2010/2010-03.html.
52. "Military Warheads as a Source of Nuclear Fuel," World Nuclear Association, updated November 2013, http://www.world-nuclear.org/info/Nuclear-Fuel-Cycle/Uranium-Resources/Military-Warheads-as-a-Source-of-Nuclear-Fuel/.
53. Uranium mining results from all countries and for the last few years, including 2010, are summarized in "World Uranium Mining Production," World Nuclear Association, updated June 2013, http://www.world-nuclear.org/info/inf23.html. The review *Forty Years of Uranium Resources, Production and Demand in Perspective: The Red Book Retrospective* (OECD, 2006) can be found at the OECD bookshop, online at http://www.oecdbookshop.org. The 2009 edition of the Red Book can also be found at the OECD online bookshop, as can more recent editions. The IAEA World Distribution of Uranium Deposits Database (UDEPO) can be found at http://infcis.iaea.org/UDEPO/About.cshtml. A print version of the 2009 status—"World Distribution of Uranium Deposits (UDEPO) with Uranium Deposit Classification" (IAEA, 2009)—can be found at http://www-pub.iaea.org/MTCD/publications/PDF/te_1629_web.pdf.
54. For an expanded analysis of this subject, please see the results of Michael Dittmar's original study at "The End of Cheap Uranium," updated June 21, 2011, http://arxiv.org/abs/1106.3617.
55. The WNA mining forecast estimate can be seen in the figure at the end of "Uranium Markets," World Nuclear Association, updated June 2013, http://www.world-nuclear.org/info/inf22.html.
56. The Energy Watch Group 2006 uranium mining forecast can be found in "Uranium Resources and Nuclear Energy," EWG, December 2006, http://www.energywatchgroup.org/fileadmin/global/pdf/EWG_Report_Uranium_3-12-2006ms.pdf.

57. The 2009 edition of the Red Book can also be found at the OECD online bookshop (http://www.oecdbookshop.org), as can more recent editions.
58. D. Eckhartt, "Nuclear Fuels for Low-Beta Fusion Reactors: Lithium Resources Revisited," *Journal of Fusion Energy* 14, no. 4 (1995): 329–41.
59. S. Featherstone, "Andrea Rossi's Black Box," *Popular Science* 281, no. 5 (2012): 62–70, 97–98.
60. R. Park, *Voodoo Science: The Road from Foolishness to Fraud* (New York: Oxford University Press, 2000).
61. J. Baez, "This Week's Finds (Week 315)," *Azimuth* (blog), June 27, 2011, http://www.johncarlosbaez.wordpress.com/2011/06/27/this-weeks-finds-week-315/.
62. D. Pimentel, C. Harvey, P. Resosudarmo, K. Sinclair, D. Kurz, M. McNair, S. Crist, L. Shpritz, L. Fitton, R. Saffouri, and R. Blair, "Environmental and Economic Costs of Soil Erosion and Conservation Benefits," *Science* 267 (1995): 1117–23.
63. R. L. B. Hooke, "On the History of Humans as Geomorphic Agents," *Geology* 28 (2000): 843–46; and B. Wilkinson, "Humans as Geologic Agents: A Deep-Time Perspective," *Geology* 22 (2005): 161–64.
64. T. J. Brown, A. S. Walters, N. E. Idoine, R. A. Shaw, C. E. Wrighton, and T. Bide, *World Mineral Production 2006–10* (Keyworth, Nottingham, UK: British Geological Survey, 2012), http://www.bgs.ac.uk/mineralsuk/statistics/worldStatistics.html.
65. F.-W. Wellmer and J. D. Becker-Platen, "Keynote Address: Global Nonfuel Mineral Resources and Sustainability," in *Proceedings for a Workshop on Deposit Modeling, Mineral Resource Assessment, and Their Role in Sustainable Development*, ed. J. A. Briskey and K. J. Schulz, US Geological Survey Circular 1294 (USGS, 2007), pubs.usgs.gov/circ/2007/1294/paper1.html.
66. A. Yarnell, "The Many Facets of Man-Made Diamonds," *Chemical and Engineering News* 82, no. 5 (2004): 26–31.

Chapter Three: Mineral Empires

1. M. G. Kovacs, trans., *The Epic of Gilgamesh* (Stanford, Calif.: Stanford University Press, 1989), electronic edition by W. Carnahan, http://www.ancienttexts.org/library/mesopotamian/gilgamesh/.
2. A. Dollinger, "Wenamen's Journey," An Introduction to the History and Culture of Pharaonic Egypt (online), updated April 2004, http://www.reshafim.org.il/ad/egypt/wenamen.htm.
3. D. Graeber, *Debt: The First 5000 Years* (New York: Melville House Publishing, 2011).
4. For an example of just such a clay-tablet promissory note, see the sample (accession number 86.11.187) held by the Metropolitan Museum of Art at http://www.metmuseum.org/Collections/search-the-collections/30000391?rpp=20&pg=1&ao=on&ft=Assyrian&what=Clay&who=Assyrian&pos=12.
5. A. M. Innes, "What Is Money?" *Banking Law Journal*, May 1913, http://www.ces.org.za/docs/what%20is%20money.htm.

6. For a review of articles pertaining to new forms of local currencies (among other topics), see "Alternative Money Systems," http://www.newciv.org/ncn/moneyteam.html.
7. For information on the Transition Town movement, refer to the Transition Town Network, online at http://www.transitionnetwork.org.
8. Editors of *Encyclopaedia Brittanica*, "Origins of Coins," *Encyclopaedia Britannica* online, updated March 26, 2012, http://www.britannica.com/EBchecked/topic/124716/coin/15838/Origins-of-coins.
9. A. Dollinger, "Thutmose III: The Battle of Megiddo," An Introduction to the History and Culture of Pharaonic Egypt (online), updated August 2000, http://www.reshafim.org.il/ad/egypt/megiddobattle.htm.
10. S. Butler, trans., *The Iliad* by Homer, MIT Internet Classical Archive, classics.mit.edu/Homer/iliad.html, accessed December 20, 2013
11. A. T. Young and S. Du, "Did Leaving the Gold Standard Tame the Business Cycle? Evidence from NBER Reference Dates and Real GNP," *Southern Economic Journal* 76, no. 2 (2009): 310–27.
12. Gold Reserve Act of 1934: Hearings Before the Committee on Banking and Currency on S.2366, 73rd Cong. (January 19–23, 1934), accessed December 29, 2013, http://fraser.stlouisfed.org/docs/meltzer/sengol34.pdf.
13. "Europe: Speculative Stampede," *Time*, March 22, 1968, http://www.time.com/time/magazine/article/0,9171,828475,00.html; M. Phillips, "Jan. 21, 1980: The Day Gold Peaked," *Wall Street Journal*, December 28, 2010, http://blogs.wsj.com/marketbeat/2010/12/28/jan-21-1980-the-day-gold-peaked/.
14. J. Laherrère, "Peak Gold, Easier to Model than Peak Oil?" guest posts (in two parts) on the Oil Drum website, November 25 and December 9, 2009, http://europe.theoildrum.com/node/5989 and http://europe.theoildrum.com/node/5995.
15. "Investment—Why, How and Where," World Gold Council, 2012, http://www.gold.org/investment/why_how_and_where/why_invest/demand_and_supply/.
16. "Gold Flows from Britain to Switzerland Surge in H1-Macquarie," Reuters, August19, 2013, http://www.reuters.com/article/2013/08/19/gold-uk-exports-macquarie-idUSL6N0GK2M920130819.
17. G. Davies, *A History of Money: From Ancient Times to the Present Day*, 2nd ed. (Cardiff: University of Wales Press, 2002).
18. CPM Group annual statistics for silver, 2010, on the Kitco Metals website at http://www.kitco.com/charts/CPM_silver.html.
19. "Supply & Demand," the Silver Institute, 2012, http://www.silverinstitute.org/site/supply-demand/.
20. US Geological Survey, *Mineral Commodity Summaries 2011* (Reston, Va.,: USGS, 2011), http://minerals.usgs.gov/minerals/pubs/mcs/2011/mcs2011.pdf.
21. D. Bol "Material Scarcity and Its Effect on Energy Solutions," presentation at the Ninth International ASPO Conference, Brussels, 2011, http://www.aspo9.be/assets/ASPO9_Thu_28_April_Bol.pdf.

22. G. Morteani and J. P. Northover, *Prehistoric Gold in Europe: Mines, Metallurgy and Manufacture* (New York: Springer-Verlag, 1994).
23. See, for example, the US Coinage Act of April 2, 1792; full text available online from the Constitution Society, http://www.constitution.org/uslaw/coinage1792.txt.
24. G. B. Haxel, J. B. Hedrick, and G. J. Orris, *Rare Earth Elements—Critical Resources for High Technology*, US Geological Survey Fact Sheet 087-02 (Reston, Va.: USGS, 2002), http://pubs.usgs.gov/fs/2002/fs087-02/.
25. E. Zimmer, "Adding Up World Silver Stock," Seeking Alpha, June 21, 2009, http://seekingalpha.com/article/144329-adding-up-world-silver-stock.
26. D. Zurbuchen, "The Real Silver Deficit," Financial Sense Archive, May 26, 2006, http://www.financialsensearchive.com/fsu/editorials/2006/0526.html.
27. R. Cowen, "Ancient Silver and Gold," chapter 6 in *Essays on Geology, History, and People* (unpublished, updated April 1999), http://mygeologypage.ucdavis.edu/cowen/~gel115/115CH6.html; and R. Swiecki, "The Gold: Gold during the Primitive Period," Alluvial Exploration & Mining, March 2011, http://www.kanada.net/alluvial/goldPrimitive.html.
28. J. W. Allan, "Gold," *Encyclopaedia Iranica*, online edition, updated February 14, 2012, http://www.iranicaonline.org/articles/gold.
29. B. S. Strauss, "Philip II of Macedon, Athens, and Silver Mining," *Hermes* 112, no. 4 (1984): 418–27.
30. C. Bartels, *Das Erzbergwerk am Rammelsberg* (Goslar, Germany: Preussag AG Metall, 1988).
31. G. Goldenberg, "Medieval Silver Ore Mining in the Southern Black Forest, Germany. Archaeological Evidence of Structural Elements and Social Aspects," abstract in *Historical Mining Activities in the Tyrol and Adjacent Areas: Impact on Environment and Human Societies*, proceedings of Mining in European History: Special Conference of the SFB HiMAT, Innsbruck, Austria, November 12–15, 2009 (Innsbruck, Austria: Innsbruck University Press, 2009), 112, http://www.uibk.ac.at/himat/publications/meh/tagungsfuehrer-meh.pdf.
32. T. E. Graedel, *Metal Stocks in Society: Scientific Synthesis*, 2010 (Paris: United Nations Environment Programme, 2010), accessed December 23, 2013, http://www.unep.org/resourcepanel/Publications/MetalStocks/tabid/56054/Default.aspx.
33. D. R. N. Rosa and R. N. Rosa, "Copper Depletion in the Iberian Pyrite Belt: Another Indicator of Global Scarcity," *Applied Earth Science* 120, no. 1 (2011): 39–43; J. Busby, "An Even Bigger Hole," Busby Report, revised August 24, 2010, http://www.after-oil.co.uk/evenbiggerhole.htm; G. M. Mudd, "Historical Trends in Base Metal Mining: Backcasting to Understand the Sustainability of Mining," in *Proceedings of the 48th Annual Conference of Metallurgists*, Canadian Metallurgical Society, Sudbury, Ontario, August 2009; and Silver Analyst, "Peak Copper," SafeHaven, December 15, 2005, http://www.safehaven.com/article/4280/peak-copper.
34. W. Zittel, *Assessment of Fossil Fuels Availability (Task 2a) and of Key Metals Availability (Task 2b)*, Progress Report 1 for the project Feasible Futures for the

Common Good: Energy Transition Paths in a Period of Increasing Resource Scarcities (Munich: Ludwig-Bölkow-Systemtechnik, March 2012).
35. R. B. Gordon, M. Bertram, and T. E. Graedel, "Metal Stocks and Sustainability," *PNAS* 103, no. 5 (2006): 1209–14.
36. B. J. Murton, "A Global Review of Non-living Resources on the Extended Continental Shelf," *Brazilian Journal of Geophysics* 18, no. 3 (2002): 281–307.
37. D. A. Singer, V. I. Berger, and B. C. Moring, "Porphyry Copper Deposits of the World: Database, Map, and Grade and Tonnage Models," US Geological Survey Open-File Report 2005-1060, version 1.0, 2005.
38. D. A. Cranstone, *A History of Mining and Mineral Exploration in Canada and Outlook for the Future* (Ottawa: Natural Resources Canada, 2002).
39. G. M. Mudd, *The Sustainability of Mining in Australia: Key Production Trends and Their Environmental Implications for the Future*, Research Report No. RR5 (Victoria, Australia: Civil Engineering Department, Monash University and Mineral Policy Institute, October 2007).
40. Zittel, *Assessment of Fossil Fuels Availability (Task 2a) and of Key Metals Availability (Task 2b)*; and V. P. Vidal and C. C. Gonzalez, *Actualización del Estúdio Prospectivo al Año 2020 del Consumo de Energia Eléctrica en la Minería del Cobre* (Comisión Chilena del Cobre, December 12, 2011).
41. Gordon, Bertram, and Graedel, "Metal Stocks and Sustainability"; J. Laherrère, "Copper Peak," post on the Oil Drum website, March 31, 2010, http://www.theoildrum.com/node/6307; and D. L. Edelstein, *2011 Minerals Yearbook: Copper*, advance release, US Geological Survey, http://minerals.usgs.gov/minerals/pubs/commodity/copper/myb1-2011-coppe.pdf.
42. Vidal and Gonzalez, *Actualización del Estúdio Prospectivo al Año 2020 del Consumo de Energia Eléctrica en la Minería del Cobre*.
43. BCS Incorporated, *Mining Industry Energy Bandwidth Study* (US Department of Energy Industrial Technologies Program, June 2007).
44. T. G. Goonan, *Copper Recycling in the United States in 2004*, US Geological Survey Circular 1196–X (USGS, 2010).
45. L. D. Roper, "Copper Depletion including Recycling," personal web page, December 29, 2012, http://www.roperld.com/science/minerals/copper.htm.
46. C. M. Cipolla, *Guns, Sails and Empires: Technological Innovation and the Early Phases of European Expansion, 1400–1700* (Minerva Press, 1966).
47. C. Dilworth, *Too Smart for Our Own Good: The Ecological Predicament of Humankind* (New York: Cambridge University Press, 2010).
48. M. T. Klare, *Blood and Oil: The Dangers and Consequences of America's Growing Dependency on Imported Petroleum* (New York: Henry Holt & Co., 2004).
49. Z. Brzezinski, *Power and Principle: Memoirs of the National Security Adviser, 1977–1981* (New York: Farrar, Strauss and Giroux, 1985).
50. D. Yergin, *The Prize: The Epic Quest for Oil, Money, and Power* (New York: Simon & Schuster, 1991).

51. D. B. Reynolds, "Peak Oil and the Fall of the Soviet Union: Lessons on the 20th Anniversary of the Collapse," post on the Oil Drum website, May 27, 2011, http://www.theoildrum.com/node/7878.
52. U. Bardi, "Peak? What Peak? King Coal Is Coming Back," *Cassandra's Legacy* (blog), February 21, 2012, http://www.cassandralegacy.blogspot.it/2012/02/peak-what-peak-king-coal-is-coming-back.html.
53. "Middle East: Total Primary Coal Production," *Titi Tudorancea Bulletin* (online), October 21, 2010, http://www.tititudorancea.com/z/ies_middle_east_coal_production.htm.
54. "Coal Statistics," World Coal Association, http://www.worldcoal.org/resources/coal-statistics/, accessed June 15, 2012.
55. A. Gsponer and J. P. Hurni, *The Physical Principles of Thermonuclear Explosives, Inertial Confinement Fusion, and the Quest for Fourth Generation Nuclear Weapons*, 5th ed., Technical Report no. 1 (Geneva: Independent Scientific Research Institute, March 1999), originally distributed at the 1997 INESAP Conference, Shanghai, China, September 8–10, 1997.
56. J. R. White, *Terrorism and Homeland Security*, 7th ed. (Belmont, Calif.: Wadsworth, 2012).
57. D. A. Sanger, "Obama Order Sped Up Wave of Cyberattacks against Iran," *New York Times*, June 1, 2012, http://www.nytimes.com/2012/06/01/world/middleeast/obama-ordered-wave-of-cyberattacks-against-iran.html?_r=1.

Chapter Four: The Universal Mining Machine

1. J. L. Simon, *The Ultimate Resource* (Princeton, N.J.: Princeton University Press, 1981).
2. "Coal Statistics," World Coal Association, accessed December 23, 2013, http://www.worldcoal.org/resources/coal-statistics/.
3. R. U. Ayres, "On the Practical Limits of Substitution," *Ecological Economics* 61, no. 1 (2007): 115–28.
4. K. R. Rábago, A. B. Lovins, and T. E. Feiler, *Energy and Sustainable Development in the Mining and Minerals Industries*, report no. 41 for the Mining, Minerals, and Sustainable Development Project of the IIED (Snowmass, Colo.: Rocky Mountain Institute, January 2001), http://pubs.iied.org/pdfs/G00540.pdf; and H. E. Goeller and A. M. Weinberg, "The Age of Substitutability," *American Economic Review* 68, no. 6 (1976): 1–11.
5. P. Odell, "Reports of the Oil's Industry Imminent Death Have Been Greatly Exaggerated," *Guardian*, February 15, 2008, http://ww.guardian.co.uk/commentisfree/2008/feb/15/oil.climatechange.
6. B. Skinner, "Earth Resources," *PNAS* 76, no. 9 (1979): 4212–17.
7. D. J. Murphy and C. A. S. Hall, "Adjusting the Economy to the New Energy Realities of the Second Half of the Age of Oil," *Ecological Modelling* 223, no. 1 (2011): 67–71.
8. R. Taylor and Z. Gethin Damon, "Countries Where Leaded Petrol Is Possibly Still Sold for Road Use as of 17th June 2011," Lead Group, Inc., website, updated November 18, 2013, http://www.lead.org.au/fs/fst27.html.

9. J. T. Kummer, "Use of Noble Metals in Automobile Exhaust Catalysts," *Journal of Physical Chemistry* 90, no. 20 (1986): 4747–52.
10. E. Alonso, F. R. Field, and R. E. Kirchain, "A Case Study of the Availability of Platinum Group Metals for Electronics Manufacturers," in *Proceedings of the 2008 IEEE International Symposium on Electronics and the Environment* (2008), 1–6.
11. C. R. M. Rao and G. S. Reddi, "Platinum Group Metals (PGM): Occurrence, Use and Recent Trends in Their Determination," *Trends in Analytical Chemistry* 19, no. 9 (2000): 565–86.
12. US Geological Survey, *Mineral Commodity Summaries 2012* (Reston, Va.: USGS, 2012), http://minerals.usgs.gov/minerals/pubs/mcs/2012/mcs2012.pdf.
13. F. Kapteijn, S. Stegenga, N. J. J. Dekker, J. W. Bijsterbosch, and J. A. Moulijn, "Alternatives to Noble Metal Catalyst for Automotive Exhaust Purifcation," *Catalysis Today* 16 (1993): 273–87; and I. Fechete, Y. Wang, and J. C. Védrine, "The Past Present and Future of Heterogeneous Catalysis," *Catalysis Today* 189 (2012): 2–27.
14. G. Qi and W. Li, "Pt-Free LaMnO3 Based Lean Nox Trap Catalyst," *Catalysis Today* 184 (2012): 72–77.
15. M. E. Gálvez, S. Ascaso, I. Tobías, R. Moliner, and M. J. Lázaro, "Catalytic Filters for the Simultaneous Removal of Soot and Nox: Influence of the Alumina Precursor on Monolith Washcoating and Catalytic Activity," *Catalysis Today* 191 (2012): 96–105.
16. R. F. Hill and W. J. Mayer, "Radiometric Determination of Platinum and Palladium Attrition from Automotive Catalysts," *Nuclear Science, IEEE Transactions* 24, no. 6 (1977): 2549–54.
17. T. E. Graedel, J. Allwood, J.-P. Birat, B. K. Reck, S. F. Sibley, G. Sonnemann, M. Buchert, and C. Hagelüken, *Recycling Rates of Metals: A Status Report* (Paris: UNEP, 2011).
18. U. Bossel, "Does a Hydrogen Economy Make Sense?" *Proceedings of the IEEE* 94, no. 10 (2006): 1826
19. Skinner, "Earth Resources."
20. K. Deffeyes, *Beyond Oil* (New York: Hill and Wang, 2005); J. W. Storm van Leeuwen, "Uranium," part D of *Nuclear Power—The Energy Balance*, October 2007, http://www.stormsmith.nl/Media/downloads/partD.pdf.
21. "Diamond Mining in Namibia: Overview," Mbendi Information Services, http://www.mbendi.com/indy/ming/dmnd/af/na/p0005.htm, accessed April 9, 2012.
22. T. R. Ajayi, "Mineral Resources of the Ocean," paper presented at the national workshop on Ocean Data and Information Network Africa (ODINAFRICA), 2000, http://www.oceandocs.net/bitstream/1834/272/1/AJAYI.pdf.
23. P. M. Herzig, M. D. Hannington, S. D. Scott, G. Thompson, and P. A. Rona, "The Distribution of Gold and Associated Trace Elements in Modern Submarine Gossans from the TAG Hydrothermal Field, Mid-Atlantic Ridge and in Ancient Ochers from Cyprus," abstract, *Geological Society of America Abstracts with Programs* 20 (1988): A240; and T. McNulty, "Gold from the Sea," goldfever.com, 1994, http://www.goldfever.com/gold_sea.htm#Herzig91.

24. Herzig, Hannington, Scott, Thompson, and Rona, "The Distribution of Gold and Associated Trace Elements in Modern Submarine Gossans from the TAG Hydrothermal Field, Mid-Atlantic Ridge and in Ancient Ochers from Cyprus."
25. A. G. Glover and G. R. Smith, "The Deep-Sea Floor Ecosystem: Current Status and Prospects of Anthropogenic Change by the Year 2025," *Environmental Conservation* 30, no. 3 (2003): 219–41.
26. U. Bardi, "Extracting Minerals from Seawater: An Energy Analysis," *Sustainability* 2, no. 4 (2010): 980–92.
27. J. Floor Anthoni, "Oceanic Abundance of Elements," Seafriends website, 2000, 2006, http://www.seafriends.org.nz/oceano/seawater.htm, accessed March 7, 2010.
28. U. Bardi, "Extracting Minerals from Seawater."
29. Ibid.
30. K. Schwochau, (1984): "Extraction of Metals from Sea Water," in *Inorganic Chemistry*, Topics in Current Chemistry 124 (Heidelberg, Germany: Springer, 1984), 91–133; and G. Nebbia, "L'estrazione di Uranio dall'acqua di mare," August 25, 1970, http://www.aspoitalia.it/component/content/article/192.
31. U. Bardi, "Extracting Minerals from Seawater."
32. A. S. Carlsson, J. B. van Beilen, R. Möller, and D. Clayton, *Micro- and Macro-Algae: Utility for Industrial Applications*, ed. Dianna Bowles (Speen, UK: CPL Press, 2007).
33. "Plutonium," World Nuclear Association, accessed September 7, 2011, http://www.world-nuclear.org/info/inf15.html.
34. T. J. Ahrens, ed., *Global Earth Physics: A Handbook of Physical Constants* (Washington, D.C.: AGU, 1995).
35. US Geological Survey, *Mineral Commodity Summaries 2012* (Reston, Va.: USGS, 2012), http://minerals.usgs.gov/minerals/pubs/mcs/.
36. K. A. Snyder, X. G. Yang, and T. J. Miller, *Hybrid Vehicle Battery Technology: The Transition from NiMH to Li-Ion*, SAE Technical Paper (Warrendale, Penn.: SAE International, 2009).
37. D. Kushnir and B. A. Sandén, "The Time Dimension and Lithium Resource Constraints for Electric Vehicles," *Resources Policy* 37, no. 1 (2012): 93–103.
38. A. Yaksic A and J. E. Tilton, "Using the Cumulative Availability Curve to Assess the Threat of Mineral Depletion: The Case of Lithium," *Resources Policy* 34, no. 4 (2009): 185–94.
39. US Geological Survey, *Mineral Commodity Summaries 2012*.
40. Ibid.
41. US Geological Survey, *Mineral Commodity Summaries* for the years 2009, 2010, and 2011, all available at http://minerals.usgs.gov/minerals/pubs/mcs/.
42. Yaksic and Tilton, "Using the Cumulative Availability Curve to Assess the Threat of Mineral Depletion"; and W. Tahil, *The Trouble with Lithium 2: Under the Microscope* (Martainville, France: Meridian International Research, May 29, 2008), http://www.meridian-int-res.com/Projects/Lithium_Microscope.pdf.

43. D. E. Garrett, *Handbook of Lithium and Natural Calcium Chloride* (Amsterdam: Academic Press, 2004); Yaksic and Tilton, "Using the Cumulative Availability Curve to Assess the Threat of Mineral Depletion."
44. US Geological Survey, *Mineral Commodity Summaries 2012*.
45. "BP Statistical Review of World Energy June 2010," a workbook containing information from the 2010 *BP Statistical Review of World Energy*, http://www.bp.com/liveassets/bp_internet/globalbp/globalbp_uk_english/reports_and_publications/statistical_energy_review_2008/STAGING/local_assets/2010_downloads/Statistical_Review_of_World_Energy_2010.xls.
46. US Geological Survey, "Lithium: Statistics and Information," USGS Minerals Information website, http://minerals.usgs.gov/minerals/pubs/commodity/lithium/.
47. US Geological Survey, *Mineral Commodity Summaries 2012*.
48. P. W. Gruber, P. A. Medina, G. A. Keoleian, et al., "Global Lithium Availability," *Journal of Industrial Ecology* 15, no. 5 (2011): 760–75; Yaksic and Tilton, "Using the Cumulative Availability Curve to Assess the Threat of Mineral Depletion"; Kushnir and Sandén, "The Time Dimension and Lithium Resource Constraints for Electric Vehicles"; and Tahil, *The Trouble with Lithium 2*.
49. S. E. Kesler, P. W. Gruber, P. A. Medina, et al., "Global Lithium Resources: Relative Importance of Pegmatite, Brine and Other Deposits," *Ore Geology Reviews* 48 (2012): 55–69.
50. T. E. Graedel, J. Allwood, J. Birat, et al., "What Do We Know about Metal Recycling Rates?" *Journal of Industrial Ecology* 15, no. 3 (2011): 355–66.
51. Kushnir and Sandén, "The Time Dimension and Lithium Resource Constraints for Electric Vehicles."
52. R. P. Bush, "Recovery of Platinum Group Metals from High Level Radioactive Waste," *Platinum Metals Review* 35, no. 4 (1991): 202–8.
53. R. Sherr, K. T. Bainbridge, and H. H. Anderson, "Transmutation of Mercury by Fast Neutrons," *Physical Review* 60, no. 7 (1941): 473–79.
54. Z. Kolarik and E. V. Renard, "Potential Applications of Fission Platinoids in Industry," *Platinum Metals Review* 49, no. 2 (2005): 79–90.
55. H. Asai and I. Ueno, "Neutron Source Based on the Repetitive Dense Plasma Focus Model," *Fusion Engineering and Design* 7 (1988–89): 335–43.
56. T. Murphy, "Stranded Resources," Do the Math, October 25, 2011, http://www.physics.ucsd.edu/do-the-math/2011/10/stranded-resources.
57. For more information on these and other uses of zinc, refer to the International Lead & Zinc Study Group (http://www.ilzsg.org) and International Zinc Association (http://www.iza.com or http://www.zinc.org).
58. A. C. Tolcin, *2011 Minerals Yearbook: Zinc,* advance release, US Geological Survey, http://minerals.usgs.gov/minerals/pubs/commodity/zinc/myb1-2011-zinc.pdf
59. A. L. Clark, J. C. Clark, and S. Pintz, *Towards the Development of a Regulatory Framework for Polymetallic Nodule Exploitation in the Area,* ISA Technical Study, No.

11 (Kingston, Jamaica: International Seabed Authority, 2013), http://www.isa.org.jm/files/documents/EN/Pubs/TStudy11.pdf.

60. Nickel Institute, "Nickel Stocks and Flows: Where in the World Does Nickel Go?" http://www.nickelinstitute.org/Sustainability/LifeCycleManagement/NickelStocksAndFlows.aspx, accessed December 29, 2013.

Chapter Five: The Bell-Shaped Curve

1. U. Bardi, "Energy Prices and Resource Depletion: Lessons from the Case of Whaling in the Nineteenth Century," *Energy Sources, Part B: Economics, Planning, and Policy* 2, no. 3 (2007): 297–304.
2. Ibid.
3. U. Bardi, "Peak Caviar," post on the Oil Drum website, August 5, 2008, http://www.europe.theoildrum.com/node/4367.
4. T. McEvoy, "The Irish Woods since Tudor Times. Their Distribution and Exploitation," *Studies: An Irish Quarterly Review* 60, no. 238 (1971): 208–11.
5. U. Bardi, "A Distant Mirror: Ireland's Great Famine," post on the Oil Drum website, December 12, 2008, http://www.theoildrum.com/node/4498.
6. W. S. Jevons, *The Coal Question* (London: McMillan and Co., 1856).
7. H. Hotelling, "The Economics of Exhaustible Resources," *Journal of Political Economy* 39, no. 2 (1931): 137–75.
8. M. Slade, "Trends in Natural-Resource Commodity Prices: An Analysis of the Time Domain," *Journal of Environmental Economics and Management* 9 (1982): 122–37; W. D. Nordhaus, "Lethal Models 2: The Limits to Growth Revisited," *Brookings Papers on Economic Activity* 2 (1992); and H. S. Houthakker, "Are Minerals Exhaustible?" *Quarterly Review of Economics and Finance* 42 (2002): 417–21.
9. J. Simon, *The Ultimate Resource* (Princeton, N.J.: Princeton University Press, 1981).
10. E. Zimmermann, *World Resources and Industries* (New York: Harper & Brothers, 1933; 2nd rev. ed., 1951).
11. R. Solow, "Technical Change and the Aggregate Production Function," *Review of Economics and Statistics* 39, no. 3 (1957): 312–20.
12. J. E. Stiglitz, "Growth with Exhaustible Natural Resources: Efficient and Optimal Growth Paths," in *Review of Economic Studies* 41, Symposium on the Economics of Exhaustible Resources (Oxford: Oxford University Press, 1974), 123–37.
13. H. E. Daly, *Ecological Economics and Sustainable Development: Selected Essays of Herman Daly* (Northampton, Mass.: Edward Elgar, 2007).
14. Nordhaus, "Lethal Models 2."
15. R. U. Ayers and B. Warr, "Accounting for Growth: The Role of Physical Work," *Structural Change and Economic Dynamics* 16, no. 2 (2005): 181–209; and R. Kümmel, "The Impact of Energy on Industrial Growth," *Energy—The International Journal* 7 (1982): 189–203.
16. G. Hardin, "The Tragedy of the Commons," *Science* 162, no. 3859 (1968): 1243–48.

17. H. S. Gordon, "The Economic Theory of a Common-Property Resource: The Fishery," *Journal of Political Economy* 62 (1954): 124–42.
18. K. T. Frank, B. Petrie, J. S. Choi, and W. C. Leggett, "Trophic Cascades in a Formerly Cod-Dominated Ecosystem," *Science* 308, no. 5728 (2005): 1621–23.
19. A. J. Lotka, *Elements of Physical Biology* (Baltimore: Williams & Wilkins, 1925); and V. Volterra, "Variazioni e fluttuazioni del numero d'individui in specie animali conviventi," in *Memorie della Regia Accademia Nazionale dei Lincei* 6, no. 2 (1926): 31–113.
20. R. B. Gordon, M. Bertran, and T. E. Gaedel, "Metal Stocks and Sustainability," *PNAS* 103, no. 5 (2006): 1209–14.
21. M. K. Hubbert, "Nuclear Energy and the Fossil Fuels," in *Drilling and Production Practice* (America Petroleum Institute, 1956).
22. T. Kelly, D. Buckingham, C. Di Francesco, and E. K. Porter, "Historical Statistics for Mineral and Material Commodities in the United States," 2013 Version, accessed December 29, 2013, http://minerals.usgs.gov/ds/2005/140/.
23. C. A. S. Hall, "An Assessment of Several of the Historically Most Influential Theoretical Models Used in Ecology and of the Data Provided in Their Support," *Ecological Modelling* 43 (1988): 5–31.
24. C. A. S. Hall, R. Powers, and W. Schoenberg, (2008): "Peak Oil, EROI, Investments and the Economy in an Uncertain Future," in *Renewable Energy Systems: Environmental and Energetic Issues*, ed. D. Pimentel (London: Elsevier, 2008), 113–16.
25. U. Bardi, A. Lavacchi, and L. Yaxley, "Modelling EROEI and Net Energy in the Exploitation of Nonrenewable Resources," *Ecological Modelling* 223 (2011): 54–58.
26. C. Hawes and C. Reed, (2006): "Theoretical Steps towards Modelling Resilience in Complex Systems," in *Proceedings of the 6th International Conference on Computational Science and Its Applications* (Glasgow, UK, May 8–11, 2006), vol. 1, part 1 (Springer-Verlag, 2006), 644–53.
27. G. Haxel, J. Hedrick, and J. Orris, *Rare Earth Elements—Critical Resources for High Technology*, US Geological Survey Fact Sheet 087-02 (Reston, Va.: USGS, 2002).
28. V. Parsani, "China's Rare Earth Dominance," Wikinvest, http://www.wikinvest.com/wiki/China%27s_Rare_Earth_Dominance, accessed May 17, 2012.
29. "The Difference Engine: More Precious than Gold," *The Economist* (online), September 17, 2010, http://www.economist.com/blogs/babbage/2010/09/rare-earth_metals.
30. C. P. Vance, "Symbiotic Nitrogen Fixation and Phosphorus Acquisition: Plant Nutrition in a World of Declining Renewable Resources," *Plant Physiology* 127, no. 2 (2001): 390–97.
31. US Geological Survey, "Phosphate Rock: Statistics and Information," accessed December 29, 2013, http://minerals.usgs.gov/minerals/pubs/commodity/phosphate_rock/myb1-2011-phosp.pdf.
32. Vance, "Symbiotic Nitrogen Fixation and Phosphorus Acquisition."
33. US Geological Survey, "Phosphate Rock: Statistics and Information."

34. P. Déry and B. Anderson, "Peak Phosphorus," Resilience website, August 13, 2007 (originally published by Energy Bulletin), http://www.energybulletin.net/node/33164.
35. C. S. Hendrix, "Applying Hubbert Curves and Linearization to Rock Phosphate," Peterson Institute for International Economics Working Paper 11-18, November 23, 2011.
36. S. White and D. Cordell, "Peak Phosphorus: The Sequel to Peak Oil," Global Phosphorus Research Initiative, 2008, http://www.phosphorusfutures.net/peak-phosphorus.
37. U. Bardi, *The Limits to Growth Revisited* (New York: Springer, 2011).
38. J. W. Forrester, "From the Ranch to System Dynamics: An Autobiography," in *Management Laureates: A Collection of Autobiographical Essays*, ed. Arthur G. Bedeian (Greenwich, Conn.: JAI Press, 1992).
39. D. Meadows, D. J. Randers, and J. D. Meadows, D. (2004): *The Limits to Growth: The 30-Years Update* (White River Junction, Vt.: Chelsea Green Publishing Company, 2004).
40. G. M. Turner, "A Comparison of the Limits to Growth with Thirty Years of Reality," Socio-Economics and the Environment in Discussion (SEED) working paper 19 (Canberra, Australia: CSIRO Sustainable Ecosystems, June 2008).
41. U. Bardi, "The Seneca Effect: Why Decline Is Faster than Growth," *Cassandra's Legacy* (blog), August 28, 2011, http://www.cassandralegacy.blogspot.it/2011/08/seneca-effect-origins-of-collapse.html.
42. D. H. Meadows, *Leverage Points: Places to Intervene in a System* (Hartland, Vt.: Sustainability Institute, 1999), http://www.donellameadows.org/wp-content/userfiles/Leverage_Points.pdf.

Chapter Six: The Dark Side of Mining

1. "Miner's Dying Words Create Song of Tribute," *The Scotsman*, April 29, 2007, http://www.scotlandonsunday.scotsman.com/minersong/Miners-dying-words-create-song.3281044.jp.
2. "Testimony Gathered by Ashley's Mines Commission," post by Laura Del Col on the Victorian Web website, updated September 26, 2002, http://www.victorianweb.org/history/ashley.html, with commentary on text taken from *Readings in European History Since 1814*, ed. J. F. Scott and A. Baltzly (Appleton-Century-Crofts, Inc., 1930), itself reprinting excerpts from the British *Parliamentary Papers* (1842).
3. B. T. Washington, *The Man Farthest Down: A Record of Observation and Study in Europe* (Garden City, N.Y.: Doubleday, Page & Co., 1913).
4. Ibid.
5. E. H. Galeano, *Open Veins of Latin America: Five Centuries of the Pillage of a Continent* (Mexico City: Siglo Veintiuno Editores, 1973).
6. "Historical Mining Disasters," Office of Mine Safety and Health Research, accessed November 17, 2013, http://www.cdc.gov/niosh/mining/statistics/minedisasters.html.

7. "End Mountaintop Removal Coal Mining," Appalachian Voices, accessed December 27, 2013, http://appvoices.org/end-mountaintop-removal/.
8. C. N. Alpers, M. P. Hunerlach, J. T. May, and R. L. Hothem, *Mercury Contamination from Historical Gold Mining in California*, US Geological Survey Fact Sheet 2005-3015, version 1.1 (Reston, Va.: USGS, November 2005), http://pubs.usgs.gov/fs/2005/3014/.
9. "In Situ Leach (ISL) Mining of Uranium," World Nuclear Association, updated June 2012, http://www.world-nuclear.org/info/inf27.html.
10. C. Tucker, "Health Concerns of 'Fracking' Drawing Increased Attention: EPA Conducting Studies on Health Effects," *The Nation's Health* 42, no. 2 (2012): 1–14.
11. C. Kuenzer, and G. Stracher, "Geomorphology of Coal Seam Fires," *Geomorphology* 138 (2011): 209–22.
12. A. Schneider, M. A. Friedl, and D. Potere, "A New Map of Global Urban Extent from MODIS Satellite Data," *Environmental Research Letters* 4, no. 4 (2009): 044003; University of Columbia News "The Growing Urbanization of the World," accessed December 29, 2013, http://www.earth.columbia.edu/news/2005/story03-07-05.html.
13. "Land Area of Countries of the World," World by Map, accessed December 29, 2013, http://world.bymap.org/LandArea.html.
14. C. Manzo, P. Stefanini, and C. Antoniou, "Il cemento avanza, quanto suolo abbiamo consumato in dieci anni," Linkiesta, accessed May 8, 2012, http://www.linkiesta.it/consumo-suolo#ixzz1uGu4Apiy.
15. U. Bardi, "Effetto Trantor: Como ti cementifico il pianeto," Effecto Cassandra (blog), November 30, 2009, http://www.ugobardi.blogspot.com/2009/11/effetto-trantor.html.
16. W. L. Rathje and C. Murphy, *Rubbish! The Archaeology of Garbology* (Tuscon: University of Arizona Press, 2001).
17. C. E. Colten and P. N. Skinner, *The Road to Love Canal: Managing Industrial Waste before EPA* (Austin: University of Texas Press, 1996).
18. E. Andreani, "Mafia at Centre of Naples' Rubbish Mess," IOL News, January 9, 2008, http://www.iol.co.za/news/world/mafia-at-centre-of-naples-rubbish-mess-1.385229.
19. P. Connett, *The Zero Waste Solution: Untrashing the Planet One Community at a Time* (White River Junction, Vt.: Chelsea Green, 2013).
20. L. Reijnders, "Cleaner Nanotechnology and Hazard Reduction of Manufactured Nanoparticles," *Journal of Cleaner Production*, Volume 14, Issue 2, 2006, Pages 124-133.
21. J. M. Pacyna, F. Steenhuisen, S. Wilson, and E. G. Pacyna, "Global Anthropogenic Emissions of Mercury to the Atmosphere," *The Encyclopedia of Earth*, updated March 29, 2011, http://www.eoearth.org/view/article/153018/; N. Pirrone, S. Cinnirella, X. Feng, R. B. Finkelman, H. R. Friedli, J. Leaner, R. Mason, A. B. Mukherjee, G. B. Stracher, D. G. Streets, and K. Telmer, "Global Mercury Emissions to the Atmosphere from Anthropogenic and Natural Sources,"

Atmospheric Chemistry and Physics Discussions 10 (2010): 4719–52, http://www.atmos-chem-phys-discuss.net/10/4719/2010/acpd-10-4719-2010-print.pdf.

22. P. Maxson, *Mercury Flows in Europe and the World: The Impact of Decommissioned Chlor-Alkali Plants* (Brussels: European Commission Directorate General for Environment, February 2004), http://ec.europa.eu/environment/chemicals/mercury/pdf/report.pdf.
23. N. E. Selin, D. J. Jacob, R. M. Yantosca, S. Strode, L. Jaeglé, and E. M. Sunderland, "Global 3-D Land-Ocean-Atmosphere Model for Mercury: Present-Day versus Preindustrial Cycles and Anthropogenic Enrichment Factors for Deposition," *Global Biogeochemical Cycles* 22, no. 2 (2008).
24. S. Cotton, "Dimethylmercury and Mercury Poisoning," University of Bristol School of Chemistry website, accessed September 5, 2011, http://www.chm.bris.ac.uk/motm/dimethylmercury/dmmh.htm.
25. S. G. Timothy, *Minamata: Pollution and the Struggle for Democracy in Postwar Japan* (Cambridge, Mass.: Harvard University Press, 2001).
26. "Review and Analysis of the Literature on the Health Effects of Dental Amalgam," Life Sciences Research Organization website, accessed September 7, 2011, http://www.lsro.org/presentation_files/amalgam/amalgam_execsum.pdf.
27. H. Sanei, S. E. Grasby, and B. Beauchamp, "Latest Permian Mercury Anomalies," *Geology* 40, no. 1 (2011): 63–66.
28. The record of various toxic substances generated by mining and industrial activities can be found on the website of the Agency for Toxic Substances and Disease Registry at http://www.atsdr.cdc.gov.
29. U. Bardi and V. Pierini, "Declining Trends of Healthy Life Years Expectancy (HLYE) in Europe," Cornell University Library (online), http://arxiv.org/abs/1311.3799.
30. D. Archer, "Fate of Fossil Fuel CO2 in Geologic Time," *Journal of Geophysical Research* 110 (2005): C09S05.
31. World Energy Council, "Survey of Energy Resources," London, 1992, 2001, 2004, 2007, 2009, and 2010, accessed December 29, 2012, http://www.worldenergy.org/wp-content/uploads/2013/09/Complete_WER_2013_Survey.pdf.
32. W. C. J. van Rensburg, "The Relationship between Resources and Reserves: Estimates for US Coal," *Resource Policy* 8, no. 1 (1982): 53–58; and M. Höök, W. Zittel, J. Schindler, and K. Aleklett, "Global Coal Production Outlooks Based on a Logistic Model," *Fuel* 89, no. 11 (2010): 3546–58.
33. J. A. Luppens, D. C. Scott, J. E. Haacke, L. M. Osmonson, T. J. Rohrbacher, and M. S. Ellis, *Assessment of Coal Geology, Resources, and Reserves in the Gillette Coalfield, Powder River Basin, Wyoming*, US Geological Survey Open-File Report 2008-1202 (Reston, Va.: USGS, February 2, 2012), http://pubs.usgs.gov/of/2008/1202/.
34. Energy Watch Group, *Coal: Resources and Future Production*, EWG Series no. 1/2007 (EWG, March 2007), http://energywatchgroup.org/fileadmin/global/pdf/EWG_Report_Coal_10-07-2007ms.pdf.

35. W. Zittel et al., "Erneuerbare Energien und Energieeffizienz als Zentralere Beitrag zur Europaischen Energieschereheit" Report to the German federal ministry for the environment, natural resources and nuclear safety 2010, accessed December 29, 2013, http://www.bmu.de/fileadmin/bmu-import/files/pdfs/allgemein/application/pdf/um08__41_819_bf.pdf.
36. A. Glikson, "The Atmosphere's Shift of State and the Origin of Extreme Weather Events," The Conversation, September 20, 2012, http://www.theconversation.edu.au/the-atmospheres-shift-of-state-and-the-origin-of-extreme-weather-events-9285.
37. D. Archer, "Methane Hydrate Stability and Anthropogenic Climate Change," *Biogeosciences* 4 (2007): 521–44.
38. S. C. Sherwood and M. Huber, M. "An Adaptability Limit to Climate Change due to Heat Stress," *PNAS* 107, no. 21 (2010): 9552–55, http://www.pnas.org/cgi/doi/10.1073/pnas.0913352107.
39. Stocker, T. F et al. *Climate Change 2013: The Physical Science Basis:* Contribution of Working Group I to the Fifth Assessment Report of the Intergovernmental Panel on Climate Change, summary for policymakers (Cambridge University Press: Cambridge, United Kingdom and New York: NY, USA, 2013), http://www.climatechange2013.org/images/uploads/WGI_AR5_SPM_brochure.pdf.
40. R. J. Brecha, "Emission Scenarios in the Face of Fossil-Fuel Peaking," *Energy Policy* 36, no. 9 (2008): 3492–504, http://campus.udayton.edu/~physics/rjb/Articles/Emissions%20scenarios%20and%20fossil-fuel%20peaking%20-%20final.pdf; W. P. Nel and C. J. Cooper, "Implications of Fossil Fuel Constraints on Economic Growth and Global Warming," *Energy Policy* 37, no. 166 (2009), http://sites.google.com/site/willem764downloads/; P. Kharecha and J. Hansen, "Implications of 'Peak Oil' for Atmospheric CO_2 and Climate," *Global Biogeochemical Cycles* 22 (2008): GB3012; and U. Bardi, "Fire or Ice? The Role of Peak Fossil Fuels in Climate Change Scenarios," post on the Oil Drum website, March 9, 2009, http://www.theoildrum.com/node/5084.
41. M. Hook and X. Tang, "Depletion of Fossil Fuels and Anthropogenic Climate Change—A Review," *Energy Policy* 52 (2013): 797–809.
42. E. A. G. Schuur and B. Abbott, "Climate Change: High Risk of Permafrost Thaw," *Nature* 480 (2011): 32–33.
43. P. J. Crutzen, "The 'Anthropocene,'" *Journal de Physique* 12, no. 10 (2002): 1–5.
44. W. F. Ruddiman, "The Anthropogenic Greenhouse Era Began Thousands of Years Ago," *Climatic Change* 61, no. 3 (2003): 261–93; and W. F. Ruddiman, S. J. Vavrus, and J. E. Kutzbach, "A Test of the Overdue-Glaciation Hypothesis," *Quaternary Science Reviews* 24 (2005): 11.
45. A. R. Brandt, "Testing Hubbert," *Energy Policy* 35 (2007): 3074–88, doi:10.1016/j.enpol.2006.11.004; and U. Bardi and A. Lavacchi, "A Simple Interpretation of Hubbert's Model of Resource Exploitation," *Energies* 2, no. 3 (2009): 646–61, doi:10.3390/en20300646.

46. U. Bardi, A. Lavacchi, and L. Yaxley, "Modelling EROEI and Net Energy in the Exploitation of Nonrenewable Resources," *Ecological Modelling* 223 (2011): 54–58.
47. D. Francis, "Torpedo Tales," *E&P Magazine*, December 28, 2006, http://www.epmag.com/archives/oilFieldHistory/145.htm.
48. Y. Hashash and J. Javier, 2011, *Evaluation of Horizontal Directional Drilling*, Civil Engineering Studies, Illinois Center for Transportation Series no. 11-095 (Urbana: Illinois Center for Transportation, November 2011).
49. G. Pitts, "The Man Who Saw Gold in Alberta's Oil Sands," *Globe and Mail* (Toronto), August 25, 2012.
50. J. Otera, "Transesterification," *Chemical Reviews* 93, no. 4 (1993): 1449–70.
51. *World Energy Outlook 2012* (International Energy Agency, November 2012), http://www.worldenergyoutlook.org.
52. J. H. Laherrere 2012 "Mise à jour Energie, Nature et les hommes du cours 2011," October 30, 2012, http://aspofrance.viabloga.com/files/JL_Sophia2012.pdf.
53. U. Bardi, "Peak Shale Oil?" *Cassandra's Legacy* (blog) August 1, 2013, http://cassandralegacy.blogspot.ca/2013/08/peak-shale-oil-what-peak.html.
54. "Rig Count Overview and Summary," Baker Hughes, accessed December 29, 2013, http://phx.corporate-ir.net/phoenix.zhtml?c=79687&p=irol-rigcountsoverview.
55. U. Bardi, "The Shale Gas Revolution: Is It Already Over?" Club of Rome website, accessed on December 29, 2013, http://www.clubofrome.org/?p=6395.
56. D. Rogers, *Shale and Wall Street: Was the Decline in Natural Gas Prices Orchestrated?* (Energy Policy Forum, February 2013), http://shalebubble.org/wp-content/uploads/2013/02/SWS-report-FINAL.pdf.
57. U. Bardi, "Peak Shale Oil?" *Cassandra's Legacy* (blog) August 1, 2013, http://cassandralegacy.blogspot.ca/2013/08/peak-shale-oil-what-peak.html.
58. I. Pearson, P. Zeniewski, F. Gracceva, P. Zastera, C. McGlade, S. Sorrell, J. Speirs, and G. Thonhauser, *Unconventional Gas: Potential Energy Market Impacts in the European Union* (Luxembourg: Publications Office of the European Union, 2012), http://ec.europa.eu/dgs/jrc/downloads/jrc_report_2012_09_unconventional_gas.pdf.
59. J. Nocera, "How Not to Fix Climate Change," *New York Times*, February 18, 2013, http://www.nytimes.com/2013/02/19/opinion/nocera-how-not-to-fix-climate-change.html.
60. R. W. Howarth, R. Santoro, and A. Ingraffea, "Methane and the Greenhouse-Gas Footprint of Natural Gas from Shale Formations: A Letter," *Climatic Change* 106 (2011): doi:10.1007/s10584-011-0061-5.
61. R. Lovett, "Industry Challenges Study That Natural Gas 'Fracking' Adds Excessively to Greenhouse Effect: Gas Industry Dismisses Claim That Shale-Derived Natural Gas Is Worse than Coal for the Climate," *Scientific American*, April 15, 2011, http://www.scientificamerican.com/article.cfm?id=industry-challenges-study-natural-gas-excessively-adds-greenhouse-effect.
62. B. McKibben, *Eaarth* (New York: Henry Holt, 2010).

Chapter Seven: The Red Queen's Race

1. J. A. Tainter, *The Collapse of Complex Societies* (New York & Cambridge: Cambridge University Press, 2003; orig. pub. 1988).
2. D. H. Meadows, Leverage Points: Places to Intervene in a System (The Hartland, Vt.: Sustainability Institute, 1999), 145-65, http://www.donellameadows.org/wp-content/userfiles/Leverage_Points.pdf.
3. H. Hotelling, "The Economics of Exhaustible Resources," *Journal of Political Economy* 39, no. 2 (1931): 137–75.
4. H. E. Goeller and A. M. Weinberg, "Age of Substitutability: Or What Do We Do When the Mercury Runs Out," in *11th Annual Foundation Lecture on a Strategy for Resources, Eindhoven, Netherlands, 18 Sep 1975* (1975), http://www.osti.gov/scitech/servlets/purl/5045860.
5. "Reducing the Fire Hazard in Aluminum-Wired Homes," *Inspectapedia*, http://www.inspectapedia.com/aluminum/Aluminum_Wiring_Risk_Reduction.htm, accessed November 22, 2012.
6. T. Norgate and J. Rankin, "The Role of Metals in Sustainable Development," CSIRO Sustainability Papers, 2002, accessed December 29, 2013, http://www.minerals.csiro.au/sd/CSIRO_Paper_LCA_Sust.pdf.
7. M. Giampietro and K. Mayumi, *The Biofuel Delusion: The Fallacy of Large Scale Agro-Biofuels Production* (London: Earthscan, 2009).
8. J. L. Sawin, *Renewables 2013 Global Status Report* (REN21 Secretariat, Paris, France, 2013), http://www.ren21.net/REN21Activities/GlobalStatusReport.aspx.
9. M. Z. Jacobson and M. A. Delucchi, "Providing All Global Energy with Wind, Water, and Solar Power. Part I: Technologies, Energy Resources, Quantities and Areas of Infrastructure, and Materials," *Energy Policy* 39, no. 3 (March 2011): 1154–69, http://dx.doi.org/10.1016/j.enpol.2010.11.040; A. García-Olivares, J. Ballabrera-Poy, E. García-Ladona, and A. Turiel, "A Global Renewable Mix with Proven Technologies and Common Materials," *Energy Policy* 41 (February 2012): 561–74, http://dx.doi.org/10.1016/j.enpol.2011.11.018.
10. D. J. Murphy and C. A. S. Hall, "Year in Review—EROI or Energy Return on (Energy) Invested," *Annals of the N.Y. Academy of Science* 1185 (2010): 102–18.
11. M. Raugei, S. Bargiglia, and S. Ulgiati, "Life Cycle Assessment and Energy Pay-back Time of Advanced Photovoltaic Modules: CdTe and CIS Compared to Poly-Si," *Energy* 32, no. 8 (2007): 1310–18; V. Fthenakis, H. C. Kim, M. Held, M. Raugei, and J. Krones, "Update of PV Energy Payback Times and Life-Cycle Greenhouse Gas Emissions," paper presented at the 24th European Photovoltaic Solar Energy Conference, September 21–25, 2009, Hamburg, Germany; and I. Kubiszewski, C. J. Cleveland, and P. K. Endres, "Meta-Analysis of Net Energy Return for Wind Power Systems," *Renewable Energy* 35 (2010): 218–25.
12. Jacobson and Delucchi, "Providing All Global Energy with Wind, Water, and Solar Power."

13. T. J. Dijkman and R. M. J. Benders, "Comparison of Renewable Fuels Based on Their Land Use Using Energy Densities," *Renewable and Sustainable Energy Reviews* 14, no. 9 (December 2010): 3148–55, http://dx.doi.org/10.1016/j.rser.2010.07.029.
14. P. Moriarty and D. Honnery, "What Is the Global Potential for Renewable Energy?" *Renewable and Sustainable Energy Reviews* 16, no. 1 (January 2012): 244–52, http://dx.doi.org/10.1016/j.rser.2011.07.151.
15. B. C. McLellan, G. D. Corder, D. P. Giurco, and K. N. Ishihara, "Renewable Energy in the Minerals Industry: A Review of Global Potential," *Journal of Cleaner Production* 32 (September 2012): 32–44, http://dx.doi.org/10.1016/j.jclepro.2012.03.016.
16. B. H. Robinson, "E-waste: An Assessment of Global Production and Environmental Impacts," *Science of the Total Environment* 408, No. 2 (December 20, 2009): 183–191.
17. P. Connett, *The Zero Waste Solution: Untrashing the Planet One Community at a Time* (White River Junction, Vt.: Chelsea Green, 2013).
18. H. Thiel, M. V. Angel, E. J. Foell, A. L. Rice, and G. Schriever, *Environmental Risks from Large-Scale Ecological Research in the Deep Sea*, a report for the Commission of the European Communities Directorate-General for Science, Research and Development (Luxembourg: Office for Official Publications of the European Communities, 1998).
19. F.-S. Zhang and I. Hideaki, "Extraction of Metals from Municipal Solid Waste Incinerator Fly Ash by Hydrothermal Process," *Journal of Hazardous Materials* 136, no. 3 (2006): 663–70.
20. W. L. Rathje and C. Murphy, *Rubbish! The Archaeology of Garbology* (Tuscon: University of Arizona Press, 2001).
21. "Landfill Mining," Environmental Alternatives website, http://www.enviroalternatives.com/landfill.html, accessed February 27, 2012.
22. Participatory Sustainable Waste Management, accessed June 24, 2012, http://www.pswm.uvic.ca/en/welcome/about.html.
23. J. Gutberlet, *Recovering Resources—Recycling Citizenship: Urban Poverty Reduction in Latin America* (Burlington, Vt.: Ashgate, 2008).
24. A. Callari and D. F. Ruccio, "Socialism, Community, and Democracy: A Postmodern Marxian Perspective," chapter 13 in *Future Directions for Heterodox Economics*, ed. J. T. Harvey and R. F. Garnett (University of Michigan Press, 2008).
25. L. Fraisse, H. Ortiz, and M. Boulianne, *Solidarity Economy: Proposal Paper for the XXI Century*, on the website of Alliance for a Responsible, Plural, and United World, November 2001, http://www.alliance21.org/en/proposals/finals/final_ecosol_en.pdf.
26. T. Graedel, M. Buchert, B. K. Reck, and G. Sonnemann, *Assessing Mineral Resources in Society: Metal Stocks & Recycling Rates* (UNEP, 2011), http://www.unep.org/pdf/Metals_Recycling_Rates_Summary.pdf.

27. Ibid., 22.
28. Ibid.
29. Ibid., 23.
30. Ibid., 8.
31. UN-HABITAT, *Solid Waste Management in the World's Cities* (London: Earthscan, 2010).
32. M. Medina, "Solid Wastes, Poverty and the Environment in Developing Country Cities," UNU-WIDER Working Paper Series no. 2010/23, March 2010.
33. UN-HABITAT, *Solid Waste Management in the World's Cities.*
34. Medina, "Solid Wastes, Poverty and the Environment in Developing Country Cities."
35. Ibid., 7.
36. M. Renner, S. Sweeney, and J. Kubit, *Green Jobs: Towards Decent Work in a Sustainable, Low-Carbon World* (UNEP/ILO/IOE/ITUC, September 2008), http://www.unep.org/PDF/UNEPGreenJobs_report08.pdf.
37. J. C. Lima, "Workers' Cooperatives in Brazil: Autonomy versus Precariousness," *Economic and Industrial Democracy* 28, no. 4 (2007): 589–621; and J. Gutberlet, "O custo social da incineração de resíduos sólidos: Recuperação de energia em detrimento da sustentabilidade," *Revista Geográfica de América Central* 2, no. 47E (2011): 1–16.
38. M. Yellishetty, G. M. Mudd, P. G. Ranjith, and A. Tharumarajah, "Environmental Life-Cycle Comparisons of Steel Production and Recycling: Sustainability Issues, Problems and Prospects," *Environmental Science and Policy* 14, no. 6 (2011):650–663.
39. J. F. Papp, *2005 Minerals Yearbook: Recycling—Metals* (US Geological Survey, February 2007), http://minerals.usgs.gov/minerals/pubs/commodity/recycle/recycmyb05.pdf.
40. For further information on cradle-to-cradle (C2C) design, see *Cradle to Cradle: Remaking the Way We Make Things*, by W. McDonough and M. Braungart (New York: North Point Press, 2002).
41. D. J. Cordier, "Rare Earths," entry in the US Geological Survey's *Mineral Commodity Summaries 2012* (Reston, Va.: USGS, 2012), 128–29.
42. M. Frondel, P. Grosche, D. Huchtemann, A. Oberheitmann, J. Peters, G. Angerer, C. Sartorius, P. Bucholz, S. Rohling, and M. Wagner, *Trends der Angebots- und Nachfragesituation bei mineralischen Rohstoffen*, project no. 09/05 of the Federal Ministry of Economics and Energy (BMWi) (Essen and Hanover: BGR, RWI Essen, and Fraunhofer ISI, 2006), 12ff.
43. P. Foster, "Diplomatic Tensions after Japanese Arrest Chinese Fisherman," *The Telegraph*, September 8, 2010.
44. "China Blocked Exports of Rare Earth Metals to Japan, traders claim," *The Telegraph*, September 24, 2010.
45. "Waste Electrical and Electrical Equipment (WEEE)," European Commission Eurostat, http://epp.eurostat.ec.europa.eu/portal/page/portal/waste/key_waste_streams/waste_electrical_electronic_equipment_weee.

46. "Statistics on the Management of Used and End-of-Life Electronics," US Environmental Protection Agency, updated November 14, 2012, http://www.epa.gov/osw/conserve/materials/ecycling/manage.htm.
47. The Basel Convention on the Control of Transboundary Movements of Hazardous Wastes and Their Disposal went into force May 5, 1992.
48. L. Kovba, "Konflikte um strategische Rohstoffe," master's thesis (FH Ludwigshafen and CER-ETH Zürich), November 2011.
49. "Facts and figures on e-waste and recycling, 2013," Electronics Takeback Coalition, accessed December 29, 2013, http://www.electronicstakeback.com/resources/facts-and-figures/.
50. M. Buchert, A. Manhart, D. Bleher, and D. Pingel, *Recycling Critical Raw Materials from Waste Electronic Equipment* (Darmstadt: Oeko-Institut, February 24, 2012), 10–11, 30.
51. A. C. Tolcin, "Indium," entry in the US Geological Survey's *Mineral Commodity Summaries 2012* (Reston, Va.: USGS, 2012), 74–75.
52. Official Journal of the European Union L 136/3 of 29.05. 2007; Regulation (EC) No. 1907/2006.
53. T. Tsuruta, "Selective Accumulation of Light or Heavy Rare Earth Elements Using Gram-Positive Bacteria," *Colloids and Surfaces. B, Biointerfaces* 52, no. 2 (2006): 117–32.
54. F. Cherubini, S. Bargigli, and S. Ulgiati, "Life Cycle Assessment (LCA) of Waste Management Strategies: Landfilling, Sorting Plant and Incineration," *Energy* 34 (2009): 2116–23.
55. D. Orlov, *Reinventing Collapse: The Soviet Experience and American Prospects* (Gabriola Island, B.C.: New Society, 2008).
56. E. Kübler-Ross and D. Kessler, *On Grief and Grieving: Finding the Meaning of Grief through the Five Stages of Loss* (New York: Simon & Schuster, 2005).
57. D. Biello, "Negating 'Climategate,'" *Scientific American*, January 28, 2010; and N. Oreskes and E. M. Conway, *Merchants of Doubt: How a Handful of Scientists Obscured the Truth on Issues from Tobacco Smoke to Global Warming* (New York: Bloomsbury, 2010).
58. J. Hansen, *Storms of My Grandchildren* (New York: Bloomsbury, 2009).
59. Y. Sun, M. M. Joachimski, P. B. Wignall, C. Yan, Y. Chen, H. Jiang, L. Wang, and X. Lai, "Lethally Hot Temperatures during the Early Triassic Greenhouse," *Science* 338, no. 6105 (2012): 366–70, abstract, http://www.sciencemag.org/content/338/6105/366.abstract.
60. C. Stager, *Deep Future: The Next 100,000 Years of Life on Earth* (New York: Thomas Dunne Books, 2011).
61. R. C. Duncan, "Evolution, Technology, and the Natural Environment: A Unified Theory of Human History," in *Proceedings of the St. Lawrence Section ASEE Annual Meeting* (Binghamton, N.Y., 1989), 14B1-11 to 14B1-20.

62. M. K. Hubbert, "Exponential Growth as a Transient Phenomenon in Human History," paper presented before the World Wildlife Fund, Fourth International Congress, San Francisco, 1976, available online at http://www.hubbertpeak.com/hubbert/wwf1976/.
63. A. S. Carlsson, J. B. van Beilen, R. Möller, and D. Clayton, *Micro- and Macro-Algae: Utility for Industrial Applications. Outputs from the EPOBIO Project* (Speen, UK; CPL Press, 2007).
64. U. Bardi, "Post Peak Mechanized Agriculture," post on the Oil Drum website, April 28, 2009, http://www.theoildrum.com/node/4606.

Epilogue: A Mineral Eschatology

1. M. Cirkovic, "A Resource Letter on Physical Eschatology," Cornell University Library (online), arXiv:astro-ph/0211413v1, November 19, 2002, http://arxiv.org/abs/astro-ph/0211413v1.
2. B. McKibben, *Eaarth* (New York: Henry Holt, 2010).

Afterword: We Can Stop Plundering the Planet

1. http://www.oilcrashmovie.com/media/oil_depletion_protocol.pdf
2. S. Stolbert, "Bush Calls for End to Ban on Offshore Oil Drilling," *New York Times*, June 19, 2008.
3. http://www.guardian.co.uk/environment/2012/oct/10/europe-rejects-ban-arctic-oil-drilling
4. http://www.guardian.co.uk/environment/2012/oct/17/biofuels-eu-climate-impact
5. N. Oreskes and E. M. Conway, *Merchants of Doubt: How a Handful of Scientists Obscured the Truth on Issues from Tobacco Smoke to Global Warming* (New York: Bloomsbury Press, 2010), 6.

About the Author

Ugo Bardi is a member of the department of earth sciences at the University of Florence, where he teaches physical chemistry. His research interests include mineral resources, renewable energy, and system dynamics applied to economics. He is a member of the Club of Rome, of the scientific committee of the Association for the Study of Peak Oil (ASPO), and Climalteranti, a group active in climate science.

He is also founder and former president of the Italian chapter of ASPO and chief editor of *Frontiers in Energy Systems and Policy*. His articles have appeared on *The Oil Drum*, *Resilience* (formerly *The Energy Bulletin*), *Financial Sense*, and *Cassandra's Legacy*. His previous books include *The Limits to Growth Revisited*.

About the Foreword Author

Jorgen Randers is the author of *2052: A Global Forecast for the Next Forty Years*, as well as coauthor of *The Limits to Growth*, *Beyond the Limits*, and *Limits to Growth: The 30-Year Update*. He is professor of climate strategy at the BI Norwegian Business School, and was previously president of BI and deputy director general of WWF International (World Wildlife Fund). He lectures internationally on sustainable development and especially climate, and is a nonexecutive member of a number of corporate boards.

About the Contributors

Philippe Bihouix has worked for more than fifteen years as an engineer and consultant in metal-intensive sectors of the energy, chemical, telecommunication, aviation, and aerospace industries. He is coauthor of the *Quel futur pour les métaux* (*The Future of Metals*), which focuses on the scarcity of mineral resources and the technical limitations of recycling processes.

Colin J. Campbell has more than forty years of experience in the oil industry, where he has worked as a petroleum geologist, manager, and consultant. He has been employed by Oxford University, Texaco, British Petroleum, Amoco, Shenandoah Oil, Norsk Hydro, and Fina and has worked with the Bulgarian and Swedish governments. He is a trustee of the Oil Depletion Analysis Center and has authored many papers and two books, including *Resource Wars: The New Landscape of Global Conflict*.

Stefano Caporali is a postdoctoral fellow at the University of Florence. His scientific interests cover wide areas of physical chemistry—from applied aspects of electrochemistry to surface chemistry. He has contributed to more than forty journal articles and conference proceedings.

Patrick Déry is chairman of the Canada-based Groupe de recherches écologiques de La Baie (GREB), which conducts research and experimental projects in sustainable agriculture and energy technologies. One of the first scientists to have studied the phenomenon of peak phosphorus, Déry plans energy policies for various governmental agencies focusing on rural communities. For more than fifteen years he has lived and worked in an ecovillage in the province of Quebec.

Luis de Sousa is a researcher at the Public Research Centre Henri Tudor in Luxembourg. He created the first Portuguese-language website dedicated to peak oil, was a cofounder of the Portugese chapter of the Association for the Study of Peak Oil and Gas, and integrated the team that started the European branch of The Oil Drum. He writes frequently about energy and its interplay with politics and economics.

Michael Dittmar, a physicist, lectures and researches at ETH Zurich and is a senior research scientist at CERN, the European Organization for Nuclear Research. Concerns about peak oil led him to study energy and the environment. His scientific articles on nuclear fission and nuclear fusion have received wide attention.

Ian Dunlop is an engineer specializing in the interaction of corporate governance, corporate responsibility, and sustainability. Formerly a senior international oil, gas, and coal industry executive, he has chaired the Australian Coal Association and the Australian Greenhouse Office Experts Group on Emissions Trading and was CEO of the Australian Institute of Company Directors. He is a member of the Club of Rome, a member of Mikhail Gorbachev's Climate Change Task Force, and director of Safe Climate Australia. He advises internationally on climate, energy, and sustainability.

Toufic El Asmar is an international consultant at the Food and Agriculture Organization of the United Nations, where he manages the European Union–funded Seventh Framework Programme (FP7). An agronomist specializing in the economics of rural development, he has previously been a researcher at the Florence-based Institute of Biometeorology of the National Research Center of Italy; an international consultant and expert in food security and early warning for the World Meteorological Organization; and a manager for the European Commission Sixth Framework Programme (FP6). He is also peer reviewer for the *Journal of Cleaner Production* and has authored patents and scientific papers.

Jutta Gutberlet teaches geography at the University of Victoria, Canada, where she guides the Community-based Research Laboratory and is the principal investigator for numerous international projects. Her research has focused on sustainable livelihoods, resource management, and poverty reduction. She currently focuses on the multiple social-economic and environmental facets of solid waste, particularly cooperative recycling in Latin America. She recently published *Recovering Resources, Recycling Citizenship: Urban Poverty Reduction in Latin America.*

Rolf Jakobi has worked in the chemical industry for Carl Freudenberg, Dixon Resine in Italy, and BASF. He has also consulted for SRI International, formerly Stanford Research Institute, and was the CEO for ICN, the International Consulting Network. After the fall of the Iron Curtain, several management-training projects in Eastern Europe paved his way into education, and he became a professor for international management at the University of Applied Science in Hof, Bavaria, and later in Ludwigshafen. As a Swiss and German citizen he tries to foster cooperation networks. His research focuses on the linkages between chemistry, politics, and the economy, with a specific concentration on the management of raw-material resources. He studied chemical engineering at the Technical University Darmstadt and economics at the University of Distant Learning Hagen and the City University of Bellevue Washington.

Marco Pagani is a PhD in the department of agriculture and nutrition at the University of Bologna. He runs the blog Ecoalfabeta, which focuses on issues of resource depletion, renewable energy, and sustainable food production. He has contributed to the book *La vita dopo il petrolio* (Life Beyond Oil) and is currently working with a team at the University of Bologna on a book about energy waste in the food chain.

Rui Namorado Rosa is professor emeritus in physics at Evora University in Portugal and founding chair of the Portugese chapter of the Association for the Study of Peak Oil and Gas. He previously worked for the UK Atomic Energy Authority, Portuguese Nuclear Energy Board, and the Technical University of Lisbon. He is a member of the Institute of Physics in London and the European Physical Society and was a founding member of the Portuguese Physical Society. He has authored hundreds of scientific articles, technical reports, and other publications.

Jörg Schindler is the former managing director of Ludwig-Bolkow-Systemtechnik GmbH (LBST) and has also been a strategy and technology consultant for sustainable energy and transport systems. At LBST, he contributed to projects on photovoltaics and other renewable energy sources, clean road transport systems, clean fuels (such as hydrogen), and the future availability of fossil fuels. For the Energy Watch Group, he coauthored, with Werner Zittel, studies on the future supply outlook of coal, oil, and uranium. He is author and coauthor of many articles and several books on oil and mobility and a founding member, serving on the board, of the German chapter of the Association for the Study of Peak Oil and Gas.

Emilia Suomalainen is a PhD in the Industrial Ecology Group, Institute of Land Use Policies and Human Environment, at the University of Lausanne, Switzerland. She was previously the environment and sustainability project manager for the Écologie Industrielle Conseil (Industrial Ecology Council) in Paris.

Karl Wagner, a biologist by training and an environmental campaigner by profession, has over the past twenty-five years conceptualized, developed, and managed successful large environmental campaigns, nationally and internationally, for a number of international organizations, including the World Wildlife Fund, where he organized the global Living Planet Campaign and European campaigns targeting EU legislation on fisheries and chemicals (REACH), among others. He is currently the director for external relations for the Club of Rome.

Werner Zittel, executive of the Ludwig-Boelkow-Foundation, is a senior scientist at Ludwig-Bolkow-Systemtechnik GmbH in Germany. Previously he has worked at the Institute for Technical Physics in Stuttgart and the Fraunhofer Institute for Solid State Technology in Munich. He is a founding member, serving on the board, of the German chapter of the Association for the Study of Peak Oil and Gas.

Illustration and Table Credits

Chapter One

Page xx Engraving by Gustave Doré illustrating Canto XXVI of Divine Comedy, Inferno, by Dante Alighieri; Wikimedia

Figure 1.1 Kelvinsong; Wikimedia

Figure 1.2 http://meteoklima.wordpress.com

Figure 1.3 Wikimedia; www.globalwarmingart.com

Figure 1.4 Ursus; Wikimedia

Figure 1.5 ASPO; www.energybulletin.net

Figure 1.6 apfelweile; Fotolia

Chapter Two

Page 28 Alf van Beem; Wikimedia

Figure 2.1 Philippe Rekacewicz, UNEP/GRID-Arenal

Figure 2.2 Michal Maňas; Wikimedia

Figure 2.3 Dorieo21; Wikimedia

Figure 2.4 Marsyas; Wikimedia

Figure 2.5 Arthur Evans 1921; Wikimedia

Figure 2.6 M. King Hubbert 1956; www.science20.com

Figure 2.7 Stuart Staniford; http://earlywarn.blogspot.it/2012/02/oil-prices.html

Figure 2.8 Rembrandt Koppelaar; http://www.theoildrum.com/node/8936; data from *Oil & Gas Journal*

Figure 2.9 SabaeL; Wikimedia

Figure 2.10 English Wikipedia

Figure 2.11 Robert A. Rhode; www.globalwarmingart.com

Figure 2.12 Rembrandt Koppelaar; www.theoildrum.com/node/8936

Figure 2.13 Minerals UK; www.bgs.ac.uk

Table 2.1 Table developed by Michael Dittmar based on data from "World Uranium Mining Production," World Nuclear Association, updated June 2013, http://www.world-nuclear.org/info/inf23.html; *Forty Years of Uranium Resources, Production and Demand in Perspective: The Red Book Retrospective* (OECD, 2006) http://www.oecdbookshop.org; the IAEA World Distribution of Uranium Deposits Database (UDEPO) can be found at http://infcis.iaea.org/UDEPO/About.cshtml. A print version of the 2009 status—"World Distribution of Uranium Deposits (UDEPO) with Uranium Deposit Classification" (IAEA, 2009)—can be found at http://www-pub.iaea.org/MTCD/publications/PDF/te_1629_web.pdf.

Chapter Three

Page 76 Panairidde; English Wikipedia
Figure 3.1 Sandstein; Wikimedia
Figure 3.2 Yelkrokoyade; English Wikipedia
Figure 3.3 *Left*, Lydian coin and silver tetradrachma from Classical Numismatic Group, Inc., http://www.cngcoins.com; *right*, Persian daric from Delfim; Wikimedia
Figure 3.4 Fruitpunchline; Wikimedia
Figure 3.5 Johan Herman Isings, ca. 1666
Figure 3.6 Data from ASPO-Newsletter, January 2007 (Great Britain), BGR (Germany), Charbonnages de France (France)

Chapter Four

Page 112 Tim Jarrett; Wikimedia
Figure 4.1 D. H. Meadows, J. Randers, and D. L. Meadows, *Limits to Growth: The 30-Year Update* (White River Junction, Vt.: Chelsea Green, 2004).
Table 4.1 Data from US Geological Survey 2005 and Norgate, T., & Rankin, W. (2002). The role of metals in sustainable development. Green Processing Pages: 49–55. Retrieved from http://www.minerals.csiro.au/sd/CSIRO_Paper_LCA_Sust.pdf

Chapter Five

Page 142 Хрюша; Wikimedia
Figure 5.1 U. Bardi, "Energy Prices and Resource Depletion: Lessons from the Case of Whaling in the Nineteenth Century," *Energy Sources, Part B: Economics, Planning, and Policy* 2, no. 3 (2007): 297–304.
Figure 5.2 Graph developed by Marco Pagani, Stefano Caporali based on data from T. Kelly, D. Buckingham, Di Francesco, C.; Porter, E.K. (2012 accessed): *Historical Statistics for Mineral and Material Commodities in the US*, Open File Report 01-006 (U.S. Geol. Survey, Washington, DC)
Figure 5.3 Ugo Bardi
Figure 5.4 Patrick Dery, Bart Anderson; figure based on Déry, P.; Anderson, B. (2007): Peak Phosphorus. In: Energy Bulletin, www.energybulletin.net/node/33164
Figure 5.5 D. H. Meadows, J. Randers, and D. L. Meadows, *Limits to Growth: The 30-Year Update*. (White River Junction, Vt.: Chelsea Green, 2004).
Table 5.1: Table developed by Marco Pagani and Stefano Caporali based on data from T. Kelly, D. Buckingham, C. Di Francesco, E.K. Porter, (2012 accessed): *Historical Statistics for Mineral and Material Commodities in the United States*, Open File Report 01-006 (US Geological Survey, Washington, DC)

Chapter Six

Page 172 Library of Congress, Prints & Photographs Division, National Child Labor Committee Collection, [LC-DIG-nclc-01070]. Photograph by Lewis Hine. url: http://www.loc.gov/pictures/item/ncl2004000171/PP/

Figure 6.1 Didier Descouens; Wikimedia

Figure 6.2 Vladimir; English Wikipedia

Figure 6.3 Ugo Bardi; data from: http://minerals.usgs.gov/ds/2005/140/#cement

Figure 6.4 Graphs developed by Werner Zittel and Jörg Schindler based on data from World Energy Council, "Survey of Energy Resources," London, 1992, 2001, 2004, 2007, 2009, 2010.

Figure 6.5 Ugo Bardi; based on data from International Energy Agency, "World Energy Outlook 2012" http://www.worldenergyoutlook.org/ (accessed Nov 15 2012)

Figure 6.6 Robert A. Rhode; www.globalwarmingart.com

Table 6.1 "BP Statistical Review of World Energy 2013," http://www.bp.com/en/global/corporate/about-bp/energy-economics/statistical-review-of-world-energy-2013.html; Verein der Kohleimporteur, Jahresbericht 2012: Fakten und Trends 2011/2012 (Hamburg: Verein der Kohleimporteur, 2012).

Chapter Seven

Page 206 VRD; Fotolia

Figure 7.1 Ra Boe; Wikimedia

Figure 7.2 Jutta Gutberlet

Figure 7.3 Isa Fernandez; shutterstock

Figure 7.4 English Wikipedia

Index

Note: Page numbers in *italics* refer to photographs and figures; page numbers followed by *t* refer to tables.